Rocks, Roots and Rattlesnakes

A Geologist's Journal: 150 Days of Discovery on the Appalachian Trail

Craig A. Eckert

MCP BOOKS

MCP Books
2301 Lucien Way #415
Maitland, FL 32751
407.339.4217
www.millcitypress.net

© 2021 by Craig A. Eckert

Cover Photo: Standing on Clinch (Tuscarora) Sandstone overlooking Wilburn and New River Valleys near Pearisburg, Virginia

Credit: Adam Shutter Roberts.

All rights reserved solely by the author. The author guarantees all contents are original and do not infringe upon the legal rights of any other person or work. No part of this book may be reproduced in any form without the permission of the author. The views expressed in this book are not necessarily those of the publisher.

Due to the changing nature of the Internet, if there are any web addresses, links, or URLs included in this manuscript, these may have been altered and may no longer be accessible. The views and opinions shared in this book belong solely to the author and do not necessarily reflect those of the publisher. The publisher therefore disclaims responsibility for the views or opinions expressed within the work.

Paperback ISBN-13: 978-1-6628-3798-2
Ebook ISBN-13: 978-1-6628-3799-9

Contents

Foreword	v
Acknowledgements	vii
Introduction: **A Brief Geological History of the Appalachian Trail**	1

Part One: It's Only Three States

Chapter One **Georgia's Endless Mountain Wilderness**	10
Chapter Two **Climbing Higher in North Carolina**	22
Chapter Three **The Great Smoky Mountains**	33
Chapter Four **The Balds**	43
Chapter Five **Tennessee and the Road to Damascus**	58

Part Two: Endless Blue Ridge Summer

Chapter Six **Return to the Trail**	77
Chapter Seven **Mt. Rodgers, Grayson Highlands and Exploding Fish Oil**	80
Chapter Eight **Virginia's Valleys and Ridgelines**	88
Chapter Nine **Virginia's Triple Crown**	97
Chapter Ten **Breaking New Records**	105
Chapter Eleven **Raining Raccoons and Washed Out on the Priest**	109
Chapter Twelve **The Shenandoahs**	115
Chapter Thirteen **Roller Coasters, Sandy Beaches and Hang-Gliders**	125
Chapter Fourteen **North of Mason-Dixon**	131
Chapter Fifteen **Pennsylvania's Valley and Ridgeline Trails**	136

Part III: The Long, Cold Hike Home

Chapter Sixteen **Flip-Flop to Maine and the 100-Mile Wilderness**	158
Chapter Seventeen **River Crossings, Bigelows and Saddlebacks**	169
Chapter Eighteen **Navigating the Notches and the Mahoosucs**	179
Chapter Nineteen **The Whites**	188
Chapter Twenty **Leaving New Hampshire**	203

Chapter Twenty-One **The Green Mountains and the Long Trail**	208
Chapter Twenty-Two **The Berkshires**	219
Chapter Twenty-Three **Winter Arrives in Connecticut**	226
Chapter Twenty-Four **New York Highlands and Crossing the Hudson**	231
Chapter Twenty-Five **Boardwalks, Stairways to Heaven and Water Gaps**	236
Chapter Twenty-Six **Sun Up to Sun Down thru the Keystone State**	242
Epilogue	279
Appendix	282
Selected References	295
Endorsements	299

Figures and Graphs*

Figure 1: Map of Appalachian Trail from Maine to Georgia
Figure 2: View of Appalachian Trail from Maine to Georgia from space
Figure 3: Geologic Time Scale
Figure 4: Position of early continents through geologic time
Figure 5: Geologic Map of Appalachian Trail from Maine to Georgia
Figure 6: Physiographic Map of Eastern US showing Appalachian Trail from Maine to Georgia
Graph 1: Type of night stay vs. Elevation
Graph 2: Miles per day hiked vs. Elevation
Graph 3: Miles per day hiked vs. Trail hours per Day
Graph 4: Miles per day hiked vs. iPhone steps per Day
Graph 5: Stride length: Feet per step (miles x 5280)/iPhone steps vs. MPH
Graph 6: Age and rock type vs. Stride length

*All graphs found in Appendix

Foreword

This book is the story about my five-month journey hiking the Appalachian Trail (Figure 1). It is written as a series of daily logs originally transcribed from my daily journals, then embellished with additional details resulting from further recollection and research. All of the stories are true, and often contain elements of human interest relating to the many people I met on my adventure. Perhaps the most enjoyable aspect of doing such a trek is encountering so many interesting people along the way.

My goal in assembling this book was to provide entertainment and information for two main audiences. First, for those who have hiked the trail, I included stories relating to places and things all of us have seen, along with frequent observations about some of those mysterious rocks beneath our feet. For those new to the trail, or who have never hiked long distances, this is a glimpse into the wonders and excitement out there to be discovered, as told by just one of the many thousands of past thru-hikers.

As a geologist, I have added a fair amount of content pertaining to the rocks I encountered, their geologic history and relevance to the present trail. Some of the content I was already familiar with or already knew, some I relearned from review of maps and literature sources, and still some content I was learning about for the first time. I have woven those explanations and observations into stories about my daily experiences. I have also attempted to keep it fairly simple, owing to what I can only hope will be a mixed audience of readers. In addition, most of my background as a geologist has been in studying sedimentary basins, and most of the rocks encountered along the trail were, well, not sedimentary. Metamorphic and igneous rocks such as those found in the Blue Ridge and New England are far more prevalent and provided an enjoyable re-learning experience for me!

During my career, one of my pet research areas was searching for evidence for reactivation of faults in the basement complex below the Appalachian sedimentary basin, which affected depositional patterns and later structural development throughout the Paleozoic era. This was one of my favorite topics in geology and involved the use of many types of geophysical datasets as well as massive amounts of well-log data in order to image, model, and interpret deep crustal faults in the metamorphic and igneous basement complex. I spent thousands of hours researching, studying, and mapping this phenomenon and conducted technical presentations on dozens of occasions. Those deeply buried basement rocks I had spent so much of my career studying and mapping were the same rocks under my feet throughout most of the Blue Ridge physiographic province from Georgia to Pennsylvania.

As I walked each day on nearly every imaginable kind of rock over the almost 2,200 miles, the surrounding geology was always on my mind. It is woven into the fabric of the text. However, this book is in no way an attempt to match or replace any of the existing geology guides to the trail, such as Chew's 1988 masterpiece, *Underfoot*. It is rather an attempt to infuse my story with frequent musings pertaining to the rocks I encountered each day, including a lot of interesting and useful advice that a future thru-hiker may appreciate.

In the Appendix section, I have combined an assortment of published trail data with analyses of my own data collected over my 150 days. Using Excel and its regression analytics, I looked for correlations between multiple parameters collected along the way and used graphics to highlight various aspects of my journey. The resulting graphs and charts will hopefully be of interest.

Lastly, I hope readers find my occasional meanders into "rockthink" both entertaining and informative. Seeing and touching the rocks at the surface is akin to just seeing the above-water tip of the iceberg. And, who knows, out of all the future readers of these pages, perhaps just one might be bitten by the geology bug and decide to spend five months on this same journey!

Acknowledgements

I am perhaps most indebted to my family for their support of my 5 ½ month expedition. My wife Jill's understanding of my desire to do this quest for such an extended period was much appreciated and did not go unnoticed. She held down the home fort for the duration of my absence and did so without questioning the sanity of my endeavors. My brother David and my mother were two of my greatest fans and supporters, always eager for regular updates on my progress, and showing up at the close of each leg as my welcome home party. My son Josh, his wife Vanita and my grandson Braeden were regular supporters and even helped me out with permits for my Smokies visit. My daughter Jessie, her husband Andrew and their sons Ezra and Rowan were always checking on my progress and eager to learn of my whereabouts and latest adventures. I also appreciated the support and trail visits from my friends Steve, Bob, Robert and Phil on several occasions. Numerous others whom I met along my journey proved to be inspiration for the continuation of my trek, including all those folks who are part of the Appalachian Trail support community from Georgia to Maine.

Lastly, I want to thank the seven early readers of the manuscript: Joan Crockett, David Eckert, Brandon Hussing, Dr. Robert Jacobi, Hana Jacobi, Chuck Moyer and Dr. Neil Washburn; as well as my copy editor Robert Burger. Their critical feedback and recommendations for improving the manuscript were truly appreciated.

To all of you, I give my heartfelt "Thank You."

Introduction

A Brief Geological History of the Appalachian Trail

It is satisfying to be able to look at a simulated view of the Earth from space and clearly see the trace of the entire Appalachian Trail (AT) (Figure 2). The approximate 2,200-mile-long footpath extends from Mt. Katahdin, Maine to Springer Mountain, Georgia mostly following ridgelines made of tough, resistant rocks, like sandstones and granites. The entire collection of mountain chains, each having its own unique geological make up, origin and history, are collectively known as the Appalachians. These peaks and highlands are merely the roots of much greater, once more massive mountain chain created during at least four orogenies, or mountain building episodes, beginning over a billion years ago. These discreet orogenies, from oldest to youngest, are known as: Grenvillian, Taconian, Acadian, and Alleghanian (Figure 3). To understand how and why these each occurred requires a brief discussion of plate tectonics.

In 1912, Alfred Waggener, a German geophysicist and meteorologist was the first scientist to propose the continental drift hypothesis, the idea that the Earth's crust is comprised of individual slabs, or plates, that slowly move around the surface of our planet, driven by some powerful unknown forces. Originally met with much skepticism, criticism and debate, this theory was eventually embraced by the scientific community following many decades of research that yielded overwhelming evidence. The theory evolved into plate tectonics, and by the 1960s and 1970s was being taught to geology students all over the world. Even as a student in the 1970s, I can still recall being taught the "theory" of plate tectonics.

Our 4.6 billion year old planet reveals an incredible story about its origin and geological history. Nearly all the physical and geological processes involved in the creation of the Earth are cyclical, on multiple and seemingly unending time scales. There are the very short duration cycles such as tides, sunrises and sets, and longer duration multi-millennial astronomical cycles, known as Milankovitch cycles, which control glacial periods. Finally, there are hundreds of million year periods, known as Wilson cycles. These very-long-term cycles were named after Dr. John Tuzo Wilson, a famous Canadian geologist/geophysicist. Wilson cycles begin with the rifting, or breaking up of a supercontinent, which in turn forms a new ocean by seafloor spreading. The final phase is the coming back together of the fragmented continents to reform a new supercontinent, thus completing the cycle. Over the past 3.6 billion years (or Giga-annum, abbreviated as Ga) of geologic time there may have been as many as seven supercontinents, (Vaalbara, Ur, Kenorland, Nuna, Rodinia, Pannotia and Pangaea) and thus seven Wilson cycles, each fragmenting, drifting and returning together again by the process of plate tectonics. The Appalachians were formed during two of the last three Wilson cycles, involving the formation and break up of the two ancient supercontinents, Rodinia and Pangaea.

During the formation of Rodinia, much of the North American continent was already assembled but was still missing parts of its present day eastern margin. About 1.3 Ga, a long-lasting 250-million-year (or Mega-annum, abbreviated as Ma) period of mountain building events, collectively known as the Grenvillian orogeny occurred in multiple phases, accreting large swaths of continental crust known as Grenville terrane, onto the now eastern part of North America. The mountains formed during this orogeny were Himalayan in scale, and eventually eroded over time, delivering sediment into nearby shallow seas, such as the Ocoee Basin. This important sedimentary basin contains incomprehensible volumes of eroded sand-, silt-, and clay-sized particles, which would eventually become rock and comprise most of the bedrock in the Great Smokies.

The crustal plates assembled to create this supercontinent remained in place for about 250 million years, as the great mountains continued to erode. Around 750 million years ago Rodinia's plates began to diverge from one another, as deep convection currents in the highly viscous but fluid lower mantle began to drive these individual jigsaw pieces of the Earth's crust in new directions. East (present coordinates) of the newly attached Grenville terrane, a rift developed, eventually resulting in the formation of a new ocean, known as the Iapetus. A new Wilson Cycle was beginning. As these plates wandered around the early planet, the deep powerful forces that pulled them apart began to once again reunite them in fits and starts, where multiple collisions and retreats finally resulted in a reunification of Earth's last supercontinent, known as Pangaea.

During the assembly of Pangaea, three more collisions, or orogenies, occurred along the present day eastern North American continental margin. Each one of these collisions caused a crumpling (faulting and folding) of the crust and the formation of long extensive mountain chains. In addition, emplacement of igneous intrusions (plutonic rocks) and extrusions (volcanic rocks) occurred throughout these time periods. The three orogenies all took place during the Paleozoic era, and are named Taconian, Acadian and Alleghanian (Figure 4).

The Taconian occurred about 460 Ma when a volcanic island arc collided with Laurentia, resulting in the formation of a new mountain range along the present day Piedmont province, extending from present day Alabama to Newfoundland. As these mountains eroded, copious amounts of sediment (sand, silt and mud) entered surrounding low areas and were deposited in rivers, shorelines and deltas. It was at this time period that the massive pulses of Silurian sandstones were deposited, resulting in some of the most resistant bedrock and ridge forming units in the Appalachians.

After about 100 million years, another collision occurred between a microcontinent, Avalonia, and Laurentia during the Acadian orogeny roughly 380 Ma. More mountain chains formed, known as the Acadian Mountains and slowly eroded sending sediment into growing sedimentary basins to the west, such as the Appalachian, Warrior, Illinois, and Michigan Basins. These growing basins slowly depressed the existing continental crust, which had been worn down over the millennia to near flattened, eroded terrain, known as peneplains. The most significant deposit during this time was the Devonian age Catskill delta, spreading sediment for hundreds of miles westward into several actively subsiding basins.

Finally, the ancient continents of Baltica and Gondwana collided with Laurentia, closing the Iapetus Ocean and forming another mountain chain about 300 Ma. This was the final of the three Paleozoic orogenies and caused folding and thrust faulting of pre-existing rock layers, resulting in an immensely wide and long new mountain range, stretching from present-day eastern Greenland to south Texas. Each one of these orogenies caused varying degrees of either further burial, or partial exhumation of the rocks that were already deeply buried in the Piedmont and Blue Ridge areas. The tremendous heat and pressure associated with these orogenies resulted in the formation of metamorphic rocks from previous sedimentary or igneous rocks. In addition, some volcanism occurred during these collisions, resulting in extrusive igneous, or volcanic rock deposits. During relaxation or extensional phases (interorogenic periods) the crust rifted or separated in some places, allowing for deep mafic (lower mantle) molten igneous rocks to rise and flow onto the surface as flood basalts. These rocks can be seen along the AT in places such as the Shenandoahs.

After about 50 million years this supercontinent began to rift, or pull apart once again, first opening up long rift valleys in the early part of the Triassic that formed extensive lakes similar to the east African rift valleys of today. Continued rifting along the continental margin between North America and Africa soon began to open a new ocean basin we know today as the Atlantic. This began a new Wilson cycle, still going on today. As the Atlantic continued to open, there were places along our continent's margin that continued to rift or crack episodically allowing more magma to rise up from deep within the Earth and eventually cool to form the rocks we presently see at the surface.

The rocks exposed along the Appalachian Trail are the remnants of once much greater mountain chains, which over time have continued to erode to their mere roots. Over the past centuries, many thousands of well-trained geologists from all over the world have worked diligently to reconstruct the geologic history of this fascinating portion of Earth's surface we know today as the Appalachian Mountains (Figure 5).

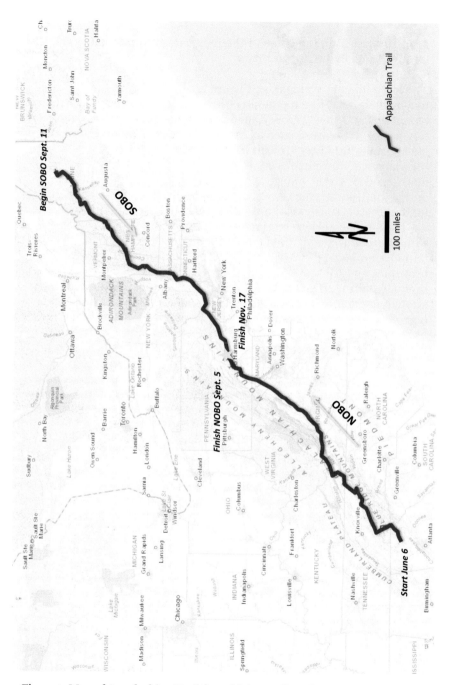

Figure 1: Map of Appalachian Trail from Maine to Georgia.
Source: National Park Service and Appalachian Trail Conservancy map app

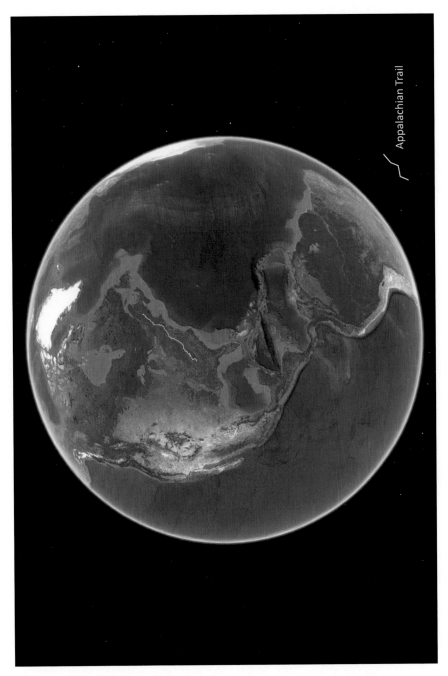

Figure 2: View of Appalachian Trail from Maine to Georgia from space
Source: Google Earth

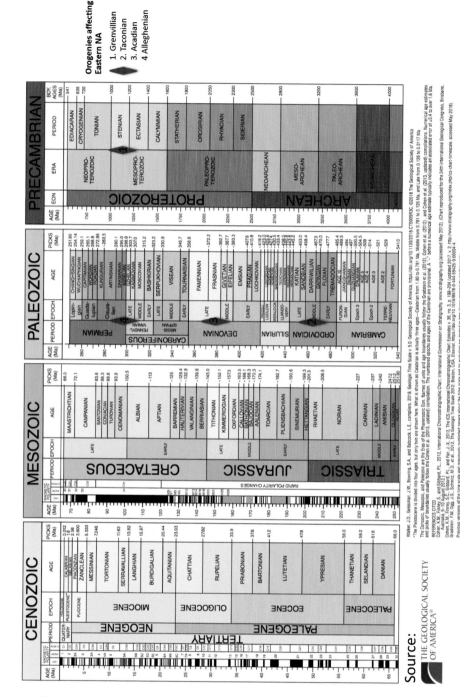

Figure 3: Geologic Time Scale. Source: Modified from Geological Society of America (Orogenic age ranges highly generalized)

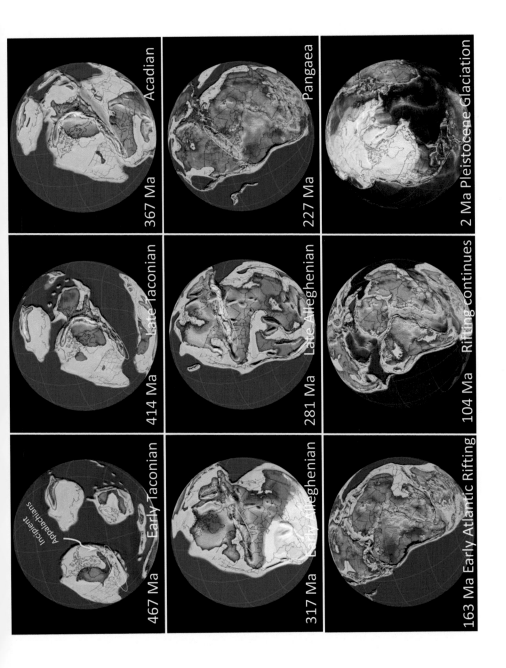

Figure 4: Position of early continents through geologic time
Source: Christopher Scotese PALEOMAP Project; GPlates

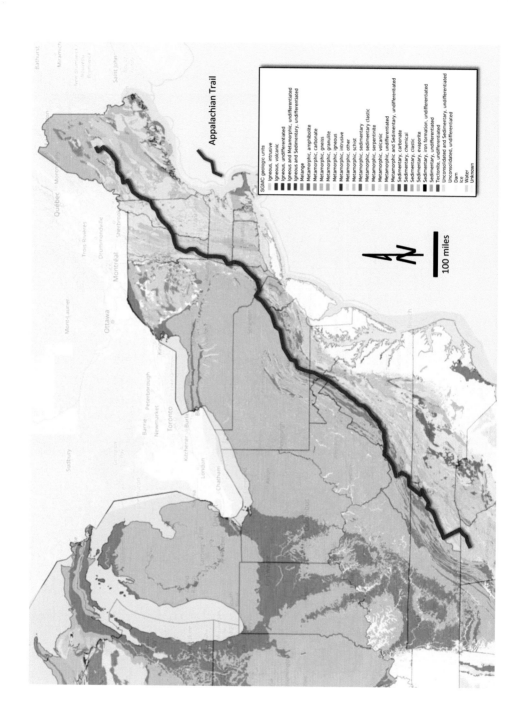

Figure 5: Geologic Map of Appalachian Trail from Maine to Georgia
Source: USGS Mineral Resources Online Spatial Data

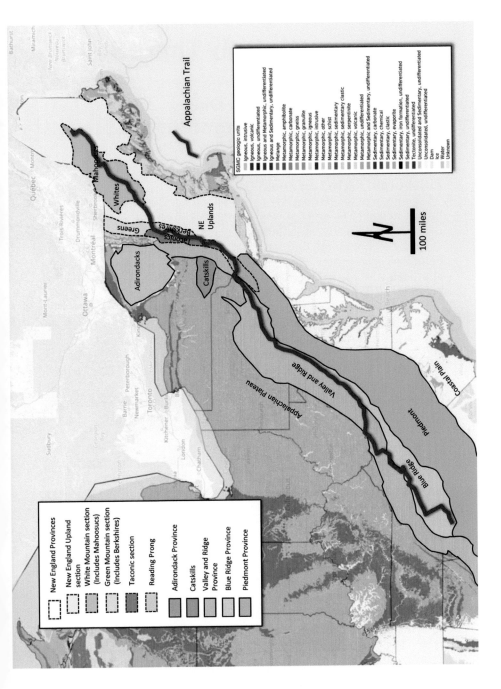

Figure 6: Physiographic Map of Eastern US showing AT from Maine to Georgia
Source: Modified from USGS Mineral Resources Online Spatial Data

Part One: It's Only Three States

Chapter One
Georgia's Endless Mountain Wilderness

Day 0 June 5, 2020 Friday
Pittsburgh, PA to Len Foote Hike Inn, GA

Today was the perfect day to start my hike. Here in the lush green hills of northern Georgia, it was clear and sunny, with temperatures in the low 80s. Already late afternoon, the time was about 5:00 PM as I stood for a photograph under the stone arch, just behind the Amicalola Falls welcome center. This iconic arch marks the beginning of the approach trail to the Appalachian Trail (AT). As I passed my starting point, there was an initial steep ascent up to a reflection pool at the base of the falls. People were fishing, sitting on benches or just strolling all around this picturesque pond. Gazing the crowd, I could see that I was the only person wearing a backpack. Everyone here seemed so relaxed and carefree, but I had this intense feeling, like I was on a mission. Continuing on from here, I began a steady climb up the 604 wooden steps to the top of the falls, all the while wearing my 50-pound L.L.Bean backpack. To say this was "not easy," would be a gross understatement. This initial stair climb would prove to be a sobering introduction of things to come over the next several months.

My weight at this point was 210 pounds, having lost about 10 pounds over the last year during my four- to eight-miles-per-day training regimen at nearby North Park, near Pittsburgh, Pennsylvania. Always without a pack on my back, my pace was a rock steady three miles per hour. I was pretty sure that I could maintain this even wearing a pack. I recall thinking that I could easily knock off 20 to 30 miles per day with no problem. I had also wondered what in the world I would do with the remainder of my daylight hours after a seven to 10 hours per day hike. How naïve I was! I would soon learn that a one- to two-mile-per-hour pace would become the norm. The realities of what I had just embarked upon were evidently still eluding me. By the end of my journey, my weight would drop to 160 pounds.

When I finally reached the top of the wooden stairs, the path next led me to a footbridge over the Amicalola River. Below me, there were impressive views of the falls cascading over the many ledges of metamorphic bedrock, accompanied by the sound of rushing water. In the far off distance, I gazed at the serenity of the surrounding hazy mountains. A solitary stick of incense that someone had left burning on one of the bridge posts gave the whole experience an ethereal feel. I continued following the blue blaze AT approach trail until about 5:40, when I reached the intersection with the green-blaze trail to the Hike Inn. My plan for the night was to stay at the Len Foote Hike Inn, a rustic lodge nestled in the Georgia Blue Ridge foothills. There is only one way to get to the Hike Inn; you literally must hike in through the Chattahoochee National Forest for about five miles. This was a moderately easy trail, and as I began my trek, I could hear the occasional low bellowing of hoot owls and the far away knocking of woodpeckers, while the path slowly wove its way through the lush green Georgia forest. A mildly fragrant assortment of wildflowers, mountain laurel, azaleas and rhododendrons were in bloom all around me. Finally, following a few very pleasant hours on this trail, I arrived at the Len Foote Hike Inn at about 8 PM.

Earlier in the day, I had arrived Atlanta at 2:16 PM, collected my 50-lb. L.L.Bean backpack in baggage claim, and met my shuttle driver Mary outside the terminal. Mary and her husband Don own a shuttle service for AT hikers, known as "The Further Shuttle." I had done some research to look for a shuttle before leaving, and Don and Mary's name were highly touted in all the reviews. We drove for two hours north to Amicalola Falls State Park Welcome Center, arriving there at 4:40 PM. I thanked Mary for the ride, and then changed into shorts and short sleeves.

Amicalola Falls State Park is in northern Georgia near the start of the AT, situated on the boundary between two physiographic provinces: the Blue Ridge to the northwest, and the Piedmont to the southeast (Figure 6). The U.S Geological Survey defines physiographic provinces as simply, "...distinctive areas having common topography, rock types and structure, and geologic and geomorphic history." All bedrock under the park is Precambrian in age, consisting primarily of metasediments, such as quartzite, metagraywacke and slate. These began as mostly sedimentary rocks, which were later deeply buried and changed into metamorphic rocks following hundreds of millions of years of intense heat, pressure, and repeated deformation the mind cannot imagine. My trek to Damascus would be almost entirely within the Blue Ridge province hiking over these very old rocks.

The original plan that I had was to just make it through the first three states of Georgia, North Carolina and Tennessee, finishing right over the Virginia border in Damascus. I had never backpacked for longer than about a week, and this hike was going to be 470 miles from Springer Mountain, Georgia to Damascus, Virginia. "But it's just three states," I thought to myself. "How hard could that really be?" The thought of all those days in the wilderness and sleeping on the ground each night was exciting and intriguing, but I wasn't sure if I could actually do it. I had recurring issues with my right knee since it was scoped for a torn meniscus about 25 years ago, following a collision with an 18-year-old opponent while attempting to head a ball during an indoor soccer match.

Also, and unrelated, two of my lumbar disks had been severely herniated and caused recurring back problems, intermittently plaguing me for at least as many years as the knee. The thought of carrying a 50-pound pack such a distance was just a bit overwhelming, not to mention sounding outright foolish. It wouldn't be until about two weeks into my adventure that I realized, "Wow, I think I can actually do this." I then said to myself, "Yes, I am going to hike the entire Appalachian Trail!" Little did I know at the time what challenges and adventures would lie ahead for me.

The air had cooled with the approach of dusk as I walked up the front wooden steps of the Hike Inn to a covered entryway and removed my backpack for the first time since I started at the visitor center. Suddenly I felt a sensation of weightlessness and had to take a few steps to regain my balance. I then entered the reception area and was greeted by Ann, who was waiting for her last arrival for the day at the front desk. She explained how everything worked with the dining room, bunkhouse and washhouse. She brought out a towel, sheets and pillowcase, and directed me to my room in the bunkhouse around the corner. I opened the screen door to a campy fragrance of a tidy cedar-walled room with bunk beds on the left and a shelf on the right. I put down my pack and took out a few items, and then ambled over to the dining hall, where the cook staff had graciously saved a dinner for my late arrival. The food was deliciously gourmet, more so than I had expected. On the surface I felt a sense of normalcy as I chatted with a family who had returned to play a board game after dinner. Inside of me I felt an inexplicable sense of nervousness and excitement.

Soon I was back in my room for the night. I remember thinking to myself, "this might be the last time I would sleep in a real bed and eat a regular meal for dinner, for perhaps a long time."

Day 0
6 miles

Day 1 June 6, 2020 Saturday
Len Foote Hike Inn, GA to Hawk Mountain Shelter, GA

At 7 AM I woke up on the top bunk in my room at the Hike Inn, where I made some coffee, and then washed up and organized my pack. This morning it was 75° and sunny, nice weather for the first day on the trail. After all, it's summer in Georgia, why wouldn't it be warm and sunny? Breakfast was not served here until 8:00 in the dining hall, so I had some time to organize and plan for my first day over a cup of hot coffee. At 8 AM, I walked over to the dining hall for a hearty breakfast of bacon, eggs and grits. I was assigned a table outside on the deck, overlooking the treetops and ridgelines of distant mountains. I met some of the kitchen staff including Matt, the general manager. Matt and the other kitchen staff would all take turns checking up with me during breakfast, occasionally offering me seconds, which of course I never turned down.

By 8:37 I was ready to go, and I left the Hike Inn on the green-blaze trail for one mile, meeting up with the blue-blaze AT approach trail. After 3.5 miles, I arrived at the summit of Springer Mountain, the start of the AT. The rocks exposed here on the ridge were of the same age and type I walked through yesterday—predominantly Precambrian metasediments. There was a plaque on a large rock there, proclaiming that this was the southern terminus of the Appalachian Trail. Next to the rock was a small metal container, inside of which was a register for all hikers to sign in before beginning their journey. I met several others at the summit; most of them were just out for the day or the weekend.

At 11 AM, I left Springer Mountain to begin my adventure. I hiked for several miles through the pristine wilderness of the Chattahoochee National Forest to Stover Creek. The trail followed the creek for several more miles, and then eventually began the lengthy ascent up Hawk Mountain passing Long Falls along the way. I could hear the sound of water gushing over rocks below me through the thick leafy curtain of green foliage.

I needed water in one of my bottles, so I got to use my nifty new Sawyer Squeeze filter for the first time. It was quick and easy; just fill up the water bag under a flowing rivulet of the stream, and then screw the filter on to the bag. Next, screw an empty one-liter Smart Water bottle to the other end of the filter and squeeze the bag gently to push it through to the empty water bottle. Simple!

The filter itself was great, but the bags, well, not so great. I would learn to despise this whole process as I filtered water day after day, with the water bags developing first one leak, then another, until only about half of the original content would pass through the filter and into my bottle, the rest squirting and dripping while I squeezed. What I finally came to realize was that the failure of the bags was totally my fault. I kept the mesh bag containing all the filter components in the rear web pocket of my pack, thus subjecting the contents to repeated daily trauma whenever I dropped my pack to the ground or against rocks. In addition, the instructions for proper use included a recommended back washing of the filter every day or so to clear the membrane of the solids accumulated during the filter process. Because I rarely attended to this important maintenance step, I had to squeeze the collection bags harder and harder in order to push water through my trusty filter, thus unduly stressing the collection bags to the point of failure.

After several miles, I arrived at Hawk Mountain Shelter. It was about 4 PM when I got there, and the faint smell of burning campfires hung in the air. There was also the indistinct sound of people talking and laughing accompanied by occasional metallic clanging of tent stakes. I noticed several campsites scattered around. At each campsite were several hikers, and even a few dogs. Some of the groups already had their own campfires started, circled by an assortment of colorful tents. Up ahead was the shelter. It reminded me of an oversized backyard shed with no front doors, set up on large rocks to keep it off the ground. There was a lot of activity and I remember thinking, "So, this is how it's going to be every day on the trail." Boy was I wrong!

    I looked for a nice, secluded area, set up my tent and then came over to the central picnic table to make my dinner. I met some of the other hikers and we exchanged idle chatter. Sitting around the fire pit were three teenage boys, who had just graduated from high school, attempting to start a campfire. They talked a lot about the civil air patrol, of which they were all proud young cadets. They were planning to hike north until they reached their home state of North Carolina. I recall one of them mentioning that he planned to hike all the way wearing a loose-fitting pair of flip-flops. I tried to imagine how this was going to work out for him. Hopefully he brought along plenty of Band-Aids and Leuko tape for his blisters.

    I also met a couple of thru-hikers, including Sunshine from Oregon, who last year completed the Pacific Crest Trail. He explained that after his completion of the AT at the end of this year, he would only need to complete the Continental Divide Trail in order to earn the "Triple Crown," North American thru-hiker's ultimate trifecta. I met another thru-hiker who was interested in checking out my tent, a Z-packs Altaplex. He came over to have a look, told me that his two-man tent, a Z-packs Duplex, was about five ounces heavier than mine, and would gladly trade with me right now if I were interested. His reasoning was solely based on conserving weight, going from a 19- to a 14-ounce tent. I told him I'd think about it, but really didn't want to make such a rash decision so early on in my trip.

    Later, I hung my food bag on one of the bear cables conveniently located right behind the shelter, and then went down to the spring to refill my water bottles. I would learn to really appreciate the convenience of bear cables and bear boxes at my camps each night. I came back to my tent and turned in for the night at about 8 PM. I had made it through my first day on the AT.

Day one
12.6 miles

Day 2 June 7, 2020 Sunday
Hawk Mountain Shelter, GA to Lance Creek Campsite, GA

    I got up at 7 AM, made breakfast and was on trail by 8:30, a late start. Today looked to be another really nice day, cool, breezy and sunny, about 65°. I hiked down the mountain to Hightower Gap, crossed over Blue Ridge Road, and while following the ridgeline, climbed up and over several small mountains to Horse Gap. The ascent up Sassafras Mountain was very steep and difficult with occasional boulder scrambles that I would become quite accustomed to in the coming months. The rocks here were mostly gneisses, schists, and other metamorphics and were ever-present along the trail. There was generally a roughness and rugosity to the gneissic bedrock, but many of the platier schists proved to be slick especially when wet. The closer I came to the ridge tops, the larger the boulders and the more actual climbing and scrambling were required to reach each summit. In reality, what I was experiencing, and what is most common on nearly all mountains whether in Georgia, Pennsylvania, or Maine, was a greater exposure of bedrock the closer to the summit I would go. This is true almost everywhere, and certainly would prove to be a predictable expectation on nearly all the summits I would climb. An important distinction should be made at this point: the presence of valley fill, rock debris, or fallen boulders should never be confused with in place (*in situ*) bedrock exposed in high mountain areas. Any large boulders found in stream valleys or New England notches have originated from higher elevations on surrounding mountains and are NOT *in situ*.

About halfway up Sassafras, I ran into a hiker who said his friend was unable to make the climb and so he was going back to his truck to pick him up at another location. I assumed that they had left a car at their planned southern endpoint and perhaps had been shuttled north for their start. I could feel the other hiker's pain, but I continued, nonetheless. Occasionally I would stop at one of the many craggy viewpoints to peer out onto the neighboring peaks and valleys, blanketed in deep summer green. However, when I finally reached the top, there were no open vistas. Instead, amongst the thick leafy foliage there were wildflowers and rhododendrons in bloom everywhere across this mountain. At one point I encountered a six-foot-long black snake stretched across my path, not at all concerned that I was standing right beside him. Everything seemed to be bursting with life as I continued following the white blazed trail, where my hike down the other side proved to be just as steep, rocky, and difficult as my ascent.

At 11 AM I arrived at Cooper Gap. From there I continued up and over Justus Mountain, to the base of Gooch Mountain, where the terrain was very easy. There was a lot of water along the way, and I even came across some impressive waterfalls. I passed by several day-hikers, and by 1:30 PM, I was at about the 8-mile mark for the day. I next passed Gooch Shelter and from there I began a very steep climb up Gooch Mountain. There were several small gaps along the way with more steep rocky inclines. After another abrupt downhill I arrived at Gooch Gap, then up and over several short but difficult climbs for the next 1.5 miles to the top of Rimrock Mountain. Here I was rewarded with a great overlook facing south into the mountainous wilderness area surrounding me.

By 3:45, I arrived at Woody Gap with pretty easy hiking for the most of the next 3.5 miles to my campground at Lance Creek. The exception was one steep rocky climb just before reaching the summit of Big Cedar Mountain. This is how it was today. Not a lot of long, slow endless ascents like I would soon experience on Blood Mountain, or in most of North Carolina, but plenty of precipitous climbs and descents covering small elevation changes of 300 to 500 feet.

While somewhere near Dan's Gap, I spotted a yellow ring-necked snake sunning itself on a flat rock in the middle of the trail. I kept plenty of these as a kid growing up in eastern Pennsylvania. My grandfather used to bring me a few snakes each time he returned from his frequent Potter County hunting trips, and I built a snake cage in my back yard where these harmless serpents could live. It was a collection of garters, red-bellies, ringed-necks, milk, and black snakes, the latter two in the family of king snakes. I soon learned that king snakes like to eat other snakes, a kind of unforeseen population control tactic in my backyard microcosm of the serpent world.

Nearing my campsite, there was the sound of water all around, as several cold mountain streams were flowing into Lance Creek. I passed a talkative young couple that had stopped to fill their water bottles but were continuing further for the night. When I finally arrived at Lance Creek there were several flat, earthen, tent platforms provided, bear cables for hanging food bags, and of course a great source of water down in the creek below. With no one else camping here on this night, I had my pick of tent sites.

Day 2
15.9 miles

Day 3 June 8, 2020 Monday
Lance Creek Campsite, GA to Baggs Creek Gap, GA

I woke up at 5:45 AM, had breakfast, broke camp, and was on trail by 7:30. This morning was cool, foggy, and drizzling rain, about 65°. My morning's hike was to the top of Blood Mountain, the highest point along the AT in Georgia, at 4459'. The ascent was steep with many rock steps and switchbacks near the top. As I neared the summit, I stopped at a few rocky overlooks to peer into the white misty abyss of the Blood Mountain Wilderness. It was very foggy and still drizzling, so it was difficult to see anything from these overlooks.

I met two section-hikers on the way up, 12-Pack and Sarah. They were also going my way and I would run into them several more times in the next day or two. Sarah explained that her trail name was Wildflower, which I thought was appropriately reflective and complimentary of her youthful exuberance, but she apparently was in search of a different one. My response when asked for my trail name was that I didn't have one yet. We continued upward together for a while, until my pace pushed me ahead. Despite the dreary weather, there were surprisingly a lot of day-hikers out here too.

All morning long I had been painfully aware of a sore left heel and arch, making it difficult and uncomfortable to hike. In general, my left heel hurt on the downhills, and my right toes hurt on the uphills. Only three days in and what a basket case I was already becoming! When I got to the top of Blood Mountain, the rain had stopped but it was still foggy, offering none of the spectacular views that I had been hearing about. I climbed to the top of some large flat-topped, house-sized boulders to rest for a bit.

The rock I was stretched out upon was mainly biotite granite and gneiss, which occasionally would glimmer in the sunlight due to the platy dark-mica flecks throughout. For most of the granites and gneisses I would encounter, the individual crystals of quartz, feldspar, or mica would be visible. This is known as phaneritic (containing visible crystals) texture, meaning the once molten rocks (magma) cooled very slowly beneath the surface in batholiths, or in some cases, in dikes, or open fractures that were later infilled with uprising magma. It is because of the slow cooling of the magma that allowed the visible crystals to form in these rocks. The longer the magma had to cool, the larger the crystals. If the magma reached the surface, as was the case during the breaking up of ancient supercontinents, like Rodinia or Pangaea, lava would erupt or flow onto the surface, thus cooling rapidly. These rocks, commonly basalts, have a very different texture owing to their rapid cooling, thus exhibiting aphanitic texture, where the individual minerals cannot be seen by the naked eye.

While at the top, I talked to several other hikers, and then climbed down to investigate the old stone shelter below. It was one of the less common AT shelters not built from wood, but rather local stone, probably because of the severe weather conditions at the top of this mountain. After a few more minutes, I began the 2.5-mile downward climb to Neel Gap. When I arrived there, the first thing I noticed was all of the old hiking boots hanging from the limbs of the surrounding trees. I went inside the outfitter store and restocked some food for the next several days, such as oatmeal, beef sticks, and a large assortment of energy bars. While there, I ran into 12-Pack and Sarah again, and we sat outside on the stone patio and talked for a while about where to camp next. I left them while they were still waiting on a pizza that they had ordered and hiked all afternoon until about 4 PM.

I decided to camp at Baggs Creek campground, just before Cowrock Mountain. There was already a man there with his tent set up, and I walked over and talked with him. He was a large man wearing a lot of camo and seemed rather shy. We walked over to get water together at the spring, and I asked him his name. He said, "They call me Slug, because I'm real slow." I told Slug I didn't have a trail name yet. Later, Sarah and 12-pack came by, set down their packs, and went for water at the spring. When they returned, they said that they had decided to continue up and over the next mountain. I wondered if my gentle giant friend, Slug, scared them off.

    I spent a lot of time getting my tent set up in preparation for what looked like a stormy night ahead. The wind was beginning to pick up, and I wanted to be sure that my trusty Altaplex would keep me dry. I then brought my cook stove and food bag over to where Slug was preparing his dinner and asked if I could join him. He turned out to be a very nice man and we talked a lot over dinner. He apparently had a very difficult earlier life, but since turned things around and now provided a halfway house for people overcoming drug and alcohol addiction. I was glad I got to know Slug, and I told him what a great and noble thing he was doing for others. Later I had a really difficult time finding a tree suitable for hanging my food bag, making me realize the value of staying at a shelter where bear cables or boxes were provided.

Day 3
11.5 miles

Day 4 June 9, 2020 Tuesday
Baggs Creek Gap, GA to Blue Mt. Shelter, GA
After waking up with first light and birds chirping, I packed up and broke camp. I was happy to find that no bears had gotten my food bag that I had so lamely hung from a low, thin branch of a nearby tree the previous night. I grabbed an energy bar and some instant coffee, then headed up to the trail toward Cowrock Mountain. The trail up the mountain was another steep ascent with lots of increasingly familiar biotite-gneiss rock steps, then down the other side to a road crossing at Tesnatee Gap. While on the summit of Cowrock, the trail opened to reveal several stunningly beautiful vistas along the ridgeline, and as I approached one of these, I noticed a familiar tent up ahead, just yards away from the cliff edge. As I passed the tent, I recognized two familiar faces gazing out toward the morning sunrise. It was 12-Pack and Sarah, apparently having made the right choice in continuing on last evening to find this perfect camping spot.

    After Tesnatee Gap, it was up and over yet another challenging ridge, then back down to Hogpen Gap. From here, the trail ascended up toward several more minor peaks but was fairly gently sloping, much of it following an old logging road. Mountain laurel and rhododendrons occupied nearly the entire trail as a canopy for miles. At one point I caught a flash of something passing by my left side, then followed by a crash of feathers and fur on a hefty tree next to me. Without hesitation, a large hawk swooped back onto a branch for a split second, only to make a second piercing dive toward a stunned squirrel, lifting the dazed rodent from the ground by his talons. With the steady pulse of his immense wings, the predator and his prey disappeared into the blur of the canopy overhead.

As I continued along the trail I noticed a pig-tailed young lady up ahead who was wearing what I would consider casual street clothes rather than hiker garb. As I approached, we both said hello, exchanged a few niceties, and then I continued onward. A while later as I was rooting through my pack for an energy bar, she caught up to me. We began talking and then started back on the trail together. She told me about her work in Chicago as a trainer for young girls in gymnastics, and about her immigration to the US from Belarus. I asked her if she was ever in the Olympics and her response was, "Yes, when I was 19. But we didn't do well that year because the Russian team was so strong." I was surprised and impressed. What are the odds of encountering an Olympic athlete on the AT? Apparently, she was on a respite from her job in Chicago and wanted to see the south. I never learned her last name, but Natasha and I hiked for the next two hours together, until she finally decided to turn back and return to her car at Hogpen Gap.

The rocks had changed to mostly quartzite and were hard on my feet as I walked along the rugged ridgeline path. It was a very gradual uphill, but because of the tough resistant rock type, there were several very abrupt, strenuous climbs to the top, and the preponderance of rocks along the trail proved to be sharp and jagged.

I met several other hikers on the trail before reaching my final destination of Blue Mountain Shelter. When I arrived, there were several hours of daylight still ahead of me. I set up my tent nearby and made dinner with some of the hikers staying there, including a very loquacious lady from West Chester, Pennsylvania who was wearing a Pitt shirt. I recall that her hammock and tarp were set up directly over an enormous puddle from the earlier rains, causing me to wonder just how in the world she could get in and out of her hammock without soaking her feet.

Day 4
14.6 miles

Day 5 June 10, 2020 Wednesday
Blue Mt. Shelter, GA to Tray Mt. Shelter, GA

Early start this morning, I broke down camp just before the rain started, and hiked down Blue Mountain. It was already feeling warm and humid. There was apparently a tropical storm in the Gulf Coast bringing storms north into the Carolinas and Georgia. The trail down the mountain was variable, and there were many very steep and difficult areas that slowed me down. Going up and over Rocky Mountain was also equally steep and treacherous in places, and just like the name implies, extremely rocky. The third mountain for the morning was Little Tray Mountain, very steep going up, and gently sloping coming down. Most of the time, I found myself looking down at the kaleidoscope of rocks and roots so that I didn't trip or stub my increasingly blistered toes.

My last ascent for the day was going up Big Tray Mountain, to an elevation of 4200 feet. As with so many of the major climbs, there were several false summits before arriving at the top, but many impressive overlooks along the way. However, because of the weather it was once again all fogged in, with heavy downpours the entire time I was making the ascent. I finally arrived at Big Tray summit in heavy rain around 2:30 PM, and although I had only logged less than 10 miles for the day, my feet were really hurting and I just needed to stop. It was a hot muggy day and although the rain was refreshing, I was happy to be able to duck under cover for a bit. Eventually the rain stopped for a short time so that I could set up my tent about 50 yards from the shelter. When the rain picked up again, I took advantage of it by showering and doing my laundry at the same time in the steady downpour (the clothes I still had on)—a bit chilly but refreshing!

A little while later, two Kentucky boys strolled up to the camp area. They were out for several days and were busily bandaging some of their very own blistered toes when I came over to greet them; this nearly ubiquitous foot care was taking trail hikers by storm, like a new ritual evolving before my eyes. They were brothers, and after introducing ourselves we talked about some of the various places in Kentucky that we all had lived. One brother lived in Louisville, the other in Lexington. I told them that I had moved our family to Flatwoods, KY from Pittsburgh in the 90's for a job in Ashland. Five years later, my job transferred me to Houston, Texas and we were very sad we had to leave Kentucky. After Texas we eventually returned to Pittsburgh, but always looked back most fondly to our time spent in the Bluegrass State.

The weather slowly improved over the course of the afternoon, but it remained very windy on top of the mountain. Fortunately, warmer temps and sun were in the forecast for tomorrow. Later that afternoon, two young ladies from Atlanta, also out for just a couple of days, joined us. I recall guessing their ages and of course as usual, I was embarrassingly way off. We all had dinner together, I eventually retired to my tent, and the four of them all stayed in the shelter for the evening.

Day 5
8.1 miles

Day 6 June 11, 2020 Thursday
Tray Mt. Shelter, GA to Campsite on Ridge Top near Little Bald Knob, GA

Early start this morning, and by 7:05 AM I was on trail. It looked like the tropical storm had finally passed. The air was still cool and foggy, but clearly beginning to dry out. I started down the mountain, there were lots of wildflowers and blooms on the laurels, rhododendron, and orange blaze azaleas. I ran into the Kentucky boys several times, early morning breakfast, later at my second breakfast stop, and then again at lunch. It was from those two that I got the idea to take off my shoes and socks for foot relief during my lunch break. This practice seemed to seriously improve my afternoon hiking comfort, and I would carry on this drill for most of the summer.

At Dick's Gap, I decided to camp early, and not try to make it all the way to Plum Orchard Shelter. This meant I would probably not see the Kentucky boys again, but my decision turned out to be a very good call. After just about a mile up the mountain, I found Little Bald Knob campsite right along the ridgeline. Situated on a long flat part of the ridge, it was warm, sunny, and breezy and I was able to hang out everything to dry on branches all around the camp area. I next followed a trail down a steep ravine to a nice cold spring where I could fill up my water bottles and wash off the day's grit and sweat.

Back at the camp, I cleaned and dried out my tent, found a comfy chair-back-shaped gneissic boulder to sit down against while I fired up my cook stove to boil some water. My dinner tonight was a Mountain House dehydrated meal packet of grilled chicken breasts and mashed potatoes. All these freeze-dried dinners were delightful and convenient luxuries to which I rarely treated myself. Tonight, I was so hungry that I hastily, and without looking at the directions, just added the boiling water to the chicken and the mashed potatoes, all at the same time. Had I read the directions, I would have taken notice to the importance of first adding water to reconstitute the meat, later followed by the powdered mashed potatoes. The result was that one of the chicken breasts never properly reconstituted. When I bit into the meat, it had the consistency of a piece of dry cork and was not very appetizing—to say the least. Another lesson learned: follow the cooking directions on the back of dehydrated food packet!

I stayed out enjoying the pleasant evening and sunset for a little while longer, and I was in my tent by 9 PM. This campsite was the first of many great "summit camps" where I would tent this summer.

Day 6
12.1 miles

Chapter Two
Climbing Higher in North Carolina

Day 7 June 12, 2020 Friday
Campsite on Ridge Top near Little Bald Knob, GA to Standing Indian Shelter, NC

Up early with first light around 5:45 AM. It was cool, breezy, and sunny with some clouds and temps about 60° in the morning, warming to 75° by afternoon. Breakfast was cold-soaked oatmeal and cold coffee; I broke camp and was on trail by 7:25 AM. I learned about "cold soaking" from one of the many "How to" YouTube videos put out there by "Darwin onthetrail," and it proved to be a good way to conserve precious cooking fuel. It essentially means just add water and wait for a little longer for the dried food to reconstitute. Typically used for foods requiring extended soaking, unlike oatmeal, but this was as far as my food experimentation had come to date.

The entire day today seemed like a continuous uphill hike. By 10 AM, I reached the Georgia-North Carolina state line. I was pretty excited about this, having completed my first state! Then on to Bly gap from there. There were lots of local day-hikers in the morning but very few in the afternoon. I ran into Rocket-T1D-73, a semi-retired NASA engineer, and his two sons, Mike (Bonanza) and Daniel (Rocky), for the second time. Rocky had been given his trail name after he developed a very bad case of poison ivy in the first week or so of their trip. His eyes were apparently so swelled shut that he was taken to a local hospital for treatment. During the ordeal, his dad and brother said he looked like Rocky Balboa, and the name stuck. They were northbound (NOBOs) but had two cars for doing 60-mile sections in reverse leapfrogging. It sounded confusing to me at first, but actually made a lot of sense and might be the most efficient way to work your way north on the trail. Because they traveled in the opposite direction in each segment, I could look forward to seeing them every few days. They had asked me my trail name, and again today I had to tell them I didn't yet have one.

So far, the AT had proven to be extremely rocky in North Carolina. These metamorphosed Proterozoic age metasediments, gneisses and occasional granites were really taking a toll on my feet and legs. Also, there were many long stretches where the trail entered a tunnel of rhododendron for a thousand feet or so. Usually there would be a dense root mat covering the trail through these "tunnels," and at this time of the year, most of these woody shrubs were in full bloom. There were also myriad wildflowers in bloom, not to mention all the countless fascinating parts of the trail, all taking my mind off the rocks and roots that I was continuously trying not to trip over. Toward the end of each day, I couldn't wait to take off my hiking shoes and slide into my dry camp shoes.

When I finally arrived at Standing Indian Shelter at about 5:30 PM, I first walked down to a spring creek to wash up and filtered water to refill my bottles. My rule of thumb at the water sources is to always go as far downstream as possible to collect water for washing, leaving the upstream pools and flowing rivulets for filling water bottles. I trusted that others followed that reasoning too. Today I drank over four liters of water, more than any other day so far. I noticed a sign posted "Shelter closed due to COVID-19," and I saw no one there or in the surrounding campsites. Later a group of two men came by but continued down the trail to another campsite. Finally, I set my tent up, unpacked, and cooked dinner by myself that evening.

Day 7
15.6 miles

Day 8 June 13, 2020 Saturday
Standing Indian Shelter, NC to Long Branch Shelter, NC
5:45 wake up to singing birds and first light. Breakfast of oatmeal and coffee, broke camp and on trail by 7 AM. I followed blue-blaze water trail to get water from the spring and prepare for my ascent up to Standing Indian Mountain. At nearly 5,500', this would be the highest mountain I had climbed since beginning the AT. The ascent was not overly severe, and I was surprised at how quickly I was able to reach the blue-blaze side trail that would lead me to the summit.

When I reached the top there were six hikers sitting around a smoldering campfire at the eastern vista point, surrounded by an assortment of tents scattered in the nearby woods. They had evidently camped here for the night and were now just relaxing and enjoying coffee while watching the sunrise. Two dogs were off to one side playing and chasing each other in circles while I sat down for a few minutes on a gneiss rock at the invitation of one of the equally nice, welcoming campers. The view from here was literally indescribable. Nestled in the heart of the Nantahalas, Standing Indian was one of the highest peaks for miles in every direction. After a brief sitting, I said goodbye and wandered out toward the edge to take some pictures. While there, I noticed that I finally had cell service and so I sent off a few texts to home.

I began the long gradual descent down the mountain, running into several groups who were on their way to the summit. The first was a family with at least four children, and one of the younger girls apparently was assigned the duty of documenting this family outing. I was able to pick her out. She was the one with the notepad and the camera. An older couple came a bit later and we stopped for a chat. The woman was a volunteer for the local trail club and she talked all about trail maintenance and what was required to keep this section of the trail free of obstructions and generally safe for hikers. The hike down also included a lot of very rooty sections of dense rhododendron thickets, which followed the many streams along the trail. When I got to Betty Creek Gap, I encountered two hikers coming towards me who warned me of a large rattlesnake on the trail up ahead. They described exactly where they had seen it, and I was on full alert. I continued up the trail towards Albert Mountain, ever vigilant, but never did see a snake.

Initially, there was a steady, gradual incline as I climbed Albert Mountain, with some parts of the trail very narrow with abrupt cliff-like drop offs down into the valley below. There were also places where I could see the trailhead washed out, and an attempt made to fashion a bridge over the washed-out portion, and then fill that in with more gravel. However, the real fun began in the last quarter-mile before reaching the fire tower. Suddenly the trail turned into a cliff climb, where trekking poles were no longer needed, and in fact, they were just in the way. I found myself scrambling up giant boulders and steep cliffs, using both hands to pull myself from one level to the next, for the last quarter mile. These rocks were tough, mainly composed of quartz-rich gneiss and quartzite, owing to their resistance and stellar ridge-forming characteristics. This small section was by far the steepest climb I had experienced to date.

When I got to the top of Albert Mountain, I considered camping there for the night, and located a nice, secluded, flat spot behind the tower and some pine trees. I left my pack there and walked over to the fire tower, built right on the exposed gneissic bedrock, where I ran into a family with two small boys. One of the young boys was very interested in rocks and asked me all kinds of questions after I informed him that I was a geologist. We had a lot of fun for a few more minutes talking about rocks as I was repeatedly approached to identify sample after sample of eroded fragments of this craggy mountaintop. I also talked to the mom and dad about my plan for a resupply, and how I was considering meeting someone at Route 64 tomorrow. The dad told me about how nice the trail was from here all the way to the highway, and that I would have no trouble hiking all the way to the next shelter within about an hour's time. He added that I could easily make it to my resupply rendezvous point by noon the next day. I decided that I would continue ahead and found his recommendation spot on.

I arrived at Long Branch Shelter about 5 o'clock where I met a family camped out there with their assorted tents occupying all but one remaining tent platform. I set up my tent on the not-so-flat earthen platform, and then came back up to the picnic area to cook my dinner. We all sat together and had a nice chat while dining. They were camped there for the night, planning to do a long loop up to Standing Indian Mountain the next day and then return to this campsite tomorrow evening.

Day 8
16.3 miles

Day 9 June 14, 2020 Sunday
Long Branch Shelter, NC to Siler Bald Shelter, NC

I awoke with sun and birds chirping, after a tough night's sleep on a sloped tent pad. This would be the last time I would sleep on such an angled slope where all night long I am clinging to the edge of my air mattress just to keep me from sliding down hill. Finding a flat tent site would be a top priority from this point on for the remainder of my backpacking adventure. I watered up and left camp around 8 AM.

It was a nice easy hike to Route 64 by noon to meet Mary with food resupply of rice, dried fruit snacks, noodles, energy bars, and fuel for my camp stove. Prior to that at about 10 AM, I came to an earlier road with a parking lot and noticed an SUV that I thought might be Mary. I walked over to check and there were two girls inside the vehicle just leaving the trail. The girl on the driver's side rolled down the window as I walked over to say hello. I told her I was sorry, but I was looking for someone else. I explained to them that I was waiting for a shuttle driver to bring me my resupply, and the car they were driving looked similar to hers. They were both very friendly and said they had some food left over from their weekend and that I was welcome to it if I would like. They gave me two dinners and several energy bars—my first trail magic! Trail magic is something hikers love, and usually involves a surprise food item or a cold drink at an unexpected time and place.

Later as I was coming down Winding Stair Gap, I noticed someone up ahead on the path with a camera. As I approached, I could see it was Mary. She had arrived early and was photographing some colorful summer fungi along the trail. It was nice to see a familiar face for a change, and we walked together down to her car in the parking area at Route 64, where she handed me a bag containing everything I asked for. I could now survive for another several days! She had even bought several large bags of gummies and candy for my afternoon snacks, which was great, and I thanked her very much for doing this for me.

I hiked most of the rest of the day in a rainstorm, the trail was flooded pretty much the whole way to my next shelter destination: Siler Bald. By 3 PM, I arrived in an intense thunderstorm with both rain and hail pelting my already soaked rain jacket. There I met Sandra, a retired nutritionist from Florida, and her high school aged grandson Tommy. Evidently the two had hiked sections of the AT together for the past several years, and they were out here this summer for their annual backpacking trip. The elevation here was 4,700' and the air temperature was already much cooler than in the lower gaps. With the addition of the icy hail falling all around, the chilled air seemed even colder, forcing each of us into our sleeping bags to warm up inside the shelter. While the cold rain continued to fall, I decided that tonight would be the first time I would not sleep in my tent.

Later that evening after 6 PM, the rain finally stopped, and the sun briefly returned for a spectacular sunset. We all hiked up the trail to Siler Bald to watch the incredible light show in the western sky as the sun gradually slumped beneath the horizon. That night, after eating ramen noodles for dinner and hanging my food bag on the bear cables, I slept for the first time, in the comfy dry interior of my wooden refuge.

Day 9
10.9 miles

Day 10 June 15, 2020 Monday
Siler Bald Shelter, NC to Copper Ridge Bald Lookout, NC

After waking up inside Siler Bald Shelter, I eventually got organized and packed up, hitting the trail by about 8 AM. It was a long muddy trudge up a steep gravely road to get back to the AT. Today was sunny and cool about 60°, at least so far. My hike this morning took me past Siler and Wine Spring Balds, and then up to 5,300' and over Wayah Bald. Around noon, I stopped for lunch and a rest on top of the grassy summit for about an hour. As I was coming down from the top, the clouds moved in, and the wind began blowing hard. I soon hit more heavy rains and lightning storms for several hours in the afternoon. I continued walking with my raincoat on and my hood up, just enjoying the sound of the rain all around in the solitude and beauty of the downpour. I began to sing some old Springsteen tunes from his first several albums, with no one in earshot.

I eventually arrived at Cold Spring Shelter about 3 PM. There I met Captain, a 10-year-old boy, and his dad Alpha, waiting out the deluge. Of course, they asked me for my trail name to which I responded, "I don't have one yet." I asked Captain if he liked candy, looking over at his dad for his nod of approval, and then proceeded to give him three large bags of candy I received yesterday. The bags were actually pretty heavy (one pound each), and I was happy to relinquish them to a 10-year-old boy as trail magic!

It was about this time that I began to realize I wasn't going to be seeing anyone often enough for them to assign me a trail name, plus I was really becoming weary of having to respond to everyone I met, "I don't have a trail name." This was day 10, and I considered several options, finally concluding that after trekking well over 100 miles so far on this footpath, I would just name myself. I thought, "How about Oriskany?" This was the formation name of a Devonian sandstone that I had spent many years of my career studying, mapping, chasing, and drilling for natural gas in the Appalachian basin. I always enjoyed working that play, and it was something I knew as much about as any geologist I might possibly meet. One of my career specialties as a petroleum geologist has always been seismic interpretation, which essentially means seeing, understanding, and accurately mapping the geology expressed in geophysical data sets acquired by seismic methods. Simply put, seismic data are acoustic signatures of complexly folded and faulted rock layers miles beneath your feet. The Oriskany sandstone was one of many target oil and gas reservoirs for which seismic interpretation was fundamental in successful exploration efforts. I finally settled on *Old Oriskany* because it seemed that nearly everyone I met out on the trail was so much younger than me.

I continued up the mountain to Copper Ridge Bald where I found another breezy ridge top campsite. I was still very wet and a bit cold, and I hung all my wet clothes on a line for the wind to dry. I quickly set up my tent and began preparing a hot dinner. While waiting for my water to boil, I decided I needed a fire to warm up to. I have always prided myself on my ability to start fires with one match and went into this challenge with my standard level of confidence. I hurriedly gathered small twigs and branches from overhanging limbs and attempted repeatedly to start a fire, but everything was just too wet. I finally gave up and returned to the business of making hot food over my gas stove. My ramen noodles seemed especially satisfying that evening, warming me from the inside as I sat huddled on a wet log in front of a cold, drenched fire ring. I planned to rise bright and early the next morning, hoping to beat a 7 AM rain forecast.

Day 10
12.7 miles

Day 11 June 16, 2020 Tuesday
Copper Ridge Bald Lookout, NC to Nantahala Outdoor Center, NC

I arose to fog and cold at over 5,000' elevation on top of Copper Ridge. I was still worried about the possibility of rain starting before I could take down my tent and break camp. I moved quickly, packed everything up, and was on the trail by 7 AM, expecting rain anytime. To my surprise the weather was sunny with abundant, puffy cumulus clouds all day, and much more pleasant than I had expected. I had great views from several more balds along the way, and much of today's trail was along open ridgelines.

There was a very steep descent down to Tellico Gap, where there was a parking lot for day-hikers to Wesser Bald observation tower. When I arrived at the gap there was a young family with two excited little girls getting ready to hike the 1.4-mile trail up to the tower. The dad kindly offered me a bottle of ice-cold water, which I gratefully accepted. I then pointed them in the direction of the AT and said I would see them at the top after I organized a few things in my pack. Even at the gap, I could see clearly to the surrounding valley and beyond, assuring me a spectacular view upon reaching the upcoming observation tower.

I followed the white blazes into the woods with my head down and the visor of my trusty Oyster Bay golf hat shading my eyes from the filtered sunlight. I began hiking uphill at a brisk pace, locomotive style, using my trekking poles for added power. Focusing on the steep and rocky terrain ahead of every step, I occasionally raised my head to look for a white blaze but did not see one anywhere. I was becoming really frustrated with the local trail clubs, and their lack of adequately maintaining these essential white blazes. After about 20 minutes of a very steep climb, up what appeared to be an old fire trail, I eventually looked up and there was the observation tower right in front of me. Evidently, I had missed the 1.4-mile-long trail and gone straight up the old fire tower road, which couldn't have been more than 0.6 mile long. I learned a valuable lesson that day, when doing a steep uphill climb, always wear my hat backwards so I can easily look up to see where I'm going!

After arriving at the observation tower, I climbed the sturdy angle-iron stairs to the top and spent about 10 minutes enjoying the 360° panoramic view while taking lots of pictures. After returning to the trail, I headed down the mountain where I again crossed paths with Rocket-T1D-73 and his two sons, Bonanza and Rocky. This was always an encounter I looked forward to every few days, and I happily informed them of the breathtaking views ahead. We exchanged several other points of interest lying ahead for both of us, and they reported that they planned to bypass the Smokies, due to the complexities of acquiring the needed permits. This came as disappointing news. We talked a while longer and then continued on our separate ways. I never did see the family from the parking lot, as they had taken the much longer, correct trail to the top.

For the remainder of the day, I hiked for about five miles down from Wesser Bald toward the Nantahala Outdoor Center (NOC), located right on the Nantahala River. Most of this trail followed totally open and exposed ridgeline providing endless views of spectacular scenery in all directions. The bedrock here was all metamorphics, mostly consisting of light mineral schists, many of which originated from the erosion of ancient Grenville mountain range formed by the assembly of the former supercontinent of Rodinia, about a billion years ago. These sediments, consisting of an assortment of shales, siltstones, and sandstones, were later buried deeply beneath mountain ranges formed throughout the Paleozoic era. These rocks were subjected to tremendous heat and pressure, and slowly transformed into the metamorphic rocks now exposed along the trail today.

The footpath was very steep and narrow in some places, and I couldn't help but notice how many fire-blackened tree stumps I passed along those miles of open knife-edge ridgeline. I was thankful that the weather cooperated on this afternoon, as it was certainly not where I would want to be during any one of the countless thunderstorms I experienced in the past two weeks. I encountered many day-hikers on my descent, including a newlywed couple from Kentucky, with the new groom hiking barefooted. This young man was the first of several barefoot hikers that I would meet on the AT, and something I still can't wrap my head around.

When I finally arrived at the NOC Village around 3 PM, the place was bustling with activity. There were tourists and hikers everywhere, rafters and kayakers floating down the river, and people zip-lining through the trees on the other side of the railroad tracks. As I followed the trail across the river, I spied a vacant Adirondack chair next to the cool and mesmerizing river, where I sat down to relax and to rest my legs and feet. After about 15 minutes, I walked next door to the outfitter store where I was greeted by lots of friendly and attentive sales workers who gladly helped me print out my permits for the Smokies. Next, I went across the street to the general store and bought some goodies to snack on, as well as my resupply for the next week. While there, I was also able to rent a room for two nights in one of the housing cabins on the other side of the railroad tracks. My room was located near the zip lines and amongst many of the young people working for the season as raft guides.

After getting into my bunkroom and showering for the first time since my stay at Amicalola Falls, I returned to village center for a juicy burger and a local draft. Tomorrow was going to be my first ZERO day for a much-needed rest!

Day 11
11.2 miles

Day 12 June 17, 2020 Wednesday
ZERO Day at Nantahala Outdoor Center, NC

I slept in until about 7:30, having been up much of the night due to eating such a heavy late dinner. In the morning I dried out and organized my gear, changed the bandages and tape on my blistered toes, and generally rested my feet and legs for tomorrow. I put my smelly camp shoes outside my door on the porch area next to a sign reading, "FREE." The shoes were too small for my feet and were even causing blisters when I walked. Within an hour, those malodorous camp shoes were gone! I later replaced them with a nice pair of water sandals I bought at the outfitter store.

Around noon, I walked over to an outdoor café, had some lunch, and while sitting out next to the swift and turbulent river, made some phone calls home and sent out lots of text messages.

Day 12
0 miles

Day 13 June 18, 2020 Thursday
Nantahala Outdoor Center, NC to Brown Fork Gap Shelter, NC

Early wake up at 5:30, outside my bunkroom it was cool, overcast and 65°. I organized my pack, bandaged up toes, and set out towards trail by 7 AM. Five hours straight up to the top of Cheoah Bald, 2,500 feet vertical up over eight miles, with only one 10-minute stop. It was a beautiful morning, and I spent a short time at the top enjoying all the great views. Looking north across the wide expanse of the Little Tennessee River valley, I could clearly see the Great Smoky Mountains, less than 25 trail miles away. I would be less than two days until I would be making my first ascent into those majestic mountains.

Due to the COVID, no thru-hiker permits for the Smokies were available in 2020. This type of permit allows thru-hikers to simply notify the park approximately when they plan to arrive but are not date and place specific. The only type of permit granted in 2020 was a backcountry permit and, for each night, only eight spots per shelter were available. When I first planned my trip, today was to be my first day in the Smokies. Clearly, I am not there yet! My son Josh, who was in Singapore with his wife's family, had kindly made my reservations for each night's stay. Unfortunately, I had overestimated my daily mileage (I had originally estimated that I could hike 15 miles per day) and now had to scrub those plans and ask Josh to make new reservations for five nights on later dates. The new reservations he made, beginning on June 20, are now consistent with my slower-than-expected pace of less than 13 miles per day.

From my lunch spot on Cheaoh Bald, it was pretty much all downhill to Stecoah Gap, then across NC State Route 143 by 4 PM, to begin the 2.5-mile marathon scramble up to Bear Fork Gap. The final miles of the day included a mild initial ascent to Sweetwater Gap, followed by an insanely steep trail for about a mile (with no water as I had run out). At the time, I thought this was one of the most difficult sections of trail yet, but I was still new here, and would over-and-over-again repeat this claim.

I finally arrived at the shelter at 6 PM, setting up my tent immediately. There were three others already there: Snapshot, Insight and Carebear. I met Snapshot when I was at the NOC, where she had so graciously given me some extra food and snack items that she received in a resupply box from home. Insight was 40 years old, had a PhD, and was legally blind. He explained to me that he had recently retired and then decided to hike the AT. I couldn't imagine how difficult it was for him to see the roots and rocks on the trail. I suppose he got a lot of help and guidance from Carebear, his female companion. After talking for a while with the other hikers, I got washed up, gathered water, made dinner, and was in my tent by 8 PM. Today had been a long and difficult day.

Day 13
16 miles

Day 14 June 19, 2020 Friday
Brown Fork Gap Shelter, NC to Fontana Dam Shelter, NC

I broke camp and was on the trail by 7:15 AM. The weather this morning was clear and cool, about 65° and partly sunny, the sky thick with those puffy cumulus clouds. Beginning about noon, those nice puffy clouds thickened and thunderstorms threatened all afternoon. The long 6.5-mile downhill from the top of Yellow River Mountain was a nice and easy trail for the most part. On my way down, I ran into Rocket-T1D-73 and his two sons coming the other way. He told me the rest of the trail down to Fontana Dam was easy and I could be there in just a couple of hours. He obviously had no idea how much my feet were killing me at the time! We talked more about the Smokies and I tried to convince them to reconsider their plans to skip them due to the permitting complexities. I left them feeling like I had maybe changed their minds and would hopefully see them in the park on our next pass by, sometime in the next several days. Later, I met several others along the trail, including Kevin from Chicago, (trail name "Tigerfrog") who was stealth camped halfway down the mountain in a small pine grove flat along the trail. A very likeable young fellow, I would see Kevin often over the following week while hiking through the Smokies.

When I arrived at the bottom, I crossed over Highway 28 and onto a Fontana Dam parking lot about 3 PM, with the sound of far-off thunder getting closer by the minute. Suddenly, the skies darkened, and it began raining really hard with lightning and thunder, so I ducked under cover at one of the shuttle stops, and waited out the rain. After 30 minutes or so of intense downpour, the storms ended, and I continued following the white blazes through the woods and along Fontana Lake toward the legendary "Fontana Hilton" Shelter area. The "Hilton" was nicknamed for its above average amenities, such as running water, solar-powered USB-recharge station and rest rooms with showers. There were several people organizing their gear or making dinner at the picnic table when I arrived. In addition, there was a good water source coming out of a faucet around back, but the rest rooms were all locked due to COVID, and the solar-powered cell-phone recharge station did not work. Big surprise!

I walked up the trail to where the tent sites were located, each one a flat concrete pad with various mechanisms built into the cement to accommodate the need for tent stakes. This was the first time I had been faced with not being able to anchor my four corner stakes where they needed to be set to ensure a proper set up of my Altaplex. Of all the positives I could name in support of my tent, one thing that was on the negative side was the need for tent stakes. Over the years, I have owned many tents, and almost every one of them was designed to be free standing, without the need for stakes. The downside with all of them of course was their weight—nowhere near the ultra-lightness of my present shelter. Anyway, I eventually figured out a way to properly anchor the corners and front guideline, and the tent was soon set up and ready to go.

After getting things unpacked and organized, I came back down, made dinner, and then inventoried my food for the next week. Tomorrow I would be entering the Smokies, with limited (if any) resupply opportunities. After tonight, I would have enough food for five more dinners, six breakfasts, five lunches, plus some assorted snacks. This was enough to get me through the Smokies, until my planned re-supply at Davenport Gap on Route 32.

Day 14
13 miles

Chapter Three
The Great Smoky Mountains

Day 15 June 20, 2020 Saturday (GSMNP Day One)
Fontana Dam Shelter, NC to Russell Field Shelter, NC

    I awoke to nice weather, with yesterday's storms finally passed. Outside I could hear the sound of birds chirping at first light, about 5:45 AM. There were low clouds over the lake, partly covering some of the mountain peaks of the Smokies. Today was going to be the day.

    The geology of the Smokies is extremely complex both from the standpoint of how the rocks were made (sedimentation, lithification, and later metamorphism), and how they were transported to where they are today (structure, involving folding and thrusting). However, in general it can be stated that most of the rocks forming the present-day mountains in the park started out as sedimentary rocks eroded more than 750 million years ago from the ancient Grenville Mountains. As these Himalayan-sized mountains, which formed over a billion years ago slowly eroded, extensive river systems carried sediment filling surrounding marine basins along the continental margin. One of these was the Ocoee Basin, to the southeast (present coordinates) along the growing rifted margin of Rodinia, where massive accumulations of sediment flowed into a shallow continental sea over a period of about 250 million years. The resulting sediment fill, known as the Ocoee Supergroup, is on the order of 10 miles in thickness. Some of these buried sedimentary rocks were subsequently further deformed and metamorphosed into metasediments such as graywacke, quartzite, and slate. Although nearly all of these rocks were subjected to some degree of metamorphism, they are still generally classified as sedimentary rocks. In addition, about 300 million years ago, during the Alleghanian Orogeny, these old pre-Cambrian rocks were thrust over top of younger sedimentary rocks along the Great Smoky Fault. These thrusted metasediments are found along much of the AT throughout the park.

    I broke camp about 7:05, noticing there was another tent set up about 20 yards away from mine. There was a young couple packing up as I walked by, and they said they were there to support a friend who had just completed an ultramarathon through the Smokies. I hiked down the road towards the dam, and in the parking lot I met the ultramarathon runner and some other friends gathered around the car snacking on fruit and other breakfast items. We chatted for a while and then I moved along. Further down the road, I ran into a hiker who informed me that someone had broken into his car and stole it from where he had parked it while he was hiking through the Smokies. He said all that was left in the parking lot was some scattered broken glass, apparently from the window broken by the car thief. He was on his way to higher ground to get a cell signal to call the police.

I continued following the white blazes down the road, to where the AT crossed the dam, and it was an impressive sight. Fontana Dam is ranked #20 of the tallest dams in the US, and #1 in the east, at 480 feet. I found it ironic that with all this electricity being generated, someone had decided to resort to solar power to provide electricity for the cell-phone recharge station back at the shelter, which incidentally did not work. At one time there used to be an incline, similar to Pittsburgh's Duquesne Incline, which carried tourists up from the Little Tennessee River to an access point where they could tour the interior of the dam. The incline and tours were discontinued following 9/11.

From the dam, it was less than a mile to the trailhead at the end of the parking lot on the other side of the dam, where there was a kiosk for submitting permits for the five nights I'd be staying at shelters. Near the trailhead, I met seven young men who said they were lost and looking for the trail to the Shuckstack Fire Tower. They were all from Nepal, and I showed them my map and pointed out the way up to the fire tower, just a short three miles on the AT.

It was a nice hike up to the top of Shuckstack Mountain, with a gentle gradient most of the way. Partway up the mountain, I noticed a pair of very expensive Ray Ban sunglasses lying on a log where the young men from Nepal had apparently stopped earlier, and I picked them up. After following a short blue-blaze trail at the ridge top, I arrived at the fire tower about 10:15 AM, where I was rewarded with great views in all directions. My Nepalese friends were busily going up and coming down the steps of the tower and I asked if anyone had lost their sunglasses. One grateful responder claimed them and thanked me for their return. I then climbed the tower, enjoyed the views, and took some pictures. Soon I was back on the northbound trail, on my way to my first shelter in the Smokies.

The trail was a steady uphill climb to Doe Knob, with several unreasonably steep parts. After climbing one of these grueling sections, I once again ran into Tigerfrog sitting next to the trail against a tree wearing a mosquito mask. We talked for a while about how far we were both planning to hike for the day, and then I continued on. My feet were really hurting for the rest of the day, and that slowed me down a lot. I finally arrived at Russell Field about 4:45. There was a family of four from Chicago milling around our wooden sanctuary, working on filtering and chemically treating their drinking water. The family consisted of dad, Brian, and two sons and a daughter; Ben, Aiden, and Emily. Brian owns a construction company in Chicago where he builds homes. They were very friendly and talkative and I really enjoyed conversing with all of them.

Later, Tigerfrog came by telling us that he had encountered a bear and had taken his picture. He also informed us of his plans to hitch a ride into Gatlinburg for a cell-phone recharge and perhaps dinner and an overnight stay at one of the countless motel options. After a short visit, he continued on down the trail. The water source here was the worst yet, and there was a sign saying to boil all water. Hopefully the sawyer squeeze filter handles that. I set my tent up in a nice flat area, and then had dinner with Brian and family. It should be a nice night, no rain in the forecast. Oh, and I never did get to see that picture of Tigerfrog's bear.

Day 15
14.9 miles

Day 16 June 21, 2020 Sunday (GSMNP Day Two)
Russell Field Shelter, NC to Silers Bald, NC

The day started out cool and overcast, but soon those low clouds opened up to reveal the sun and its warmth. I left camp early about 6:45 AM, and my first few miles went fast and were on good, even trail. Because of the many balds, or fields as they are known here, there were also many great vistas near Spence Field Shelter. Looking south I could see Little Tennessee Valley and Fontana Lake beneath the rising clouds that had earlier been nestled in the valleys below. Looking to the north afforded views of the upcoming Thunderhead and Rocky Top peaks. While gazing at the far-off horizon, I met a photographer who had come out to capture early morning photos of the surrounding mountains, and we chatted for a few minutes about photography, his camera, and the Smokies.

As I approached Rocky Top, the trail became steeper and rockier, as I expected from the name. I made it to the top and there were great views at several points along the way. There was a large boulder at the top engraved with several names including Hop Harris and Red Waldron, along with the dates from back in the 1800's when they apparently had visited the summit. Continuing along the ridgeline, I next passed four hikers who looked like they had just run a marathon. They complained about how difficult the ascent was, coming from their direction. When they asked me how the trail ahead was, I responded, "Pretty easy, I made it to the top with just a couple of minor scrambles." To that one of the hikers replied, "Yeah right, says the guy with the giant calves!" I glanced down at my calves, and they appeared pretty normal sized to me. Later as I descended Rocky Top, the trail took on the characteristics of a steep, poorly maintained drainage ditch, full of washed-out areas and odd-sized, mostly sandstone and orthoquartzite boulders. For about the next five miles, this poor trail state continued—all the way to Derrick Knob. Now I could see the other hiker's point.

Shortly after passing by Derrick Knob, the trail seemed to improve greatly. However, the wind picked up, and when I was within about 0.5 miles of Siler's Bald, heavy rains accompanied by thunder and lightning followed me the rest of the way. When I arrived at the shelter, there was a young Ukrainian lady waiting out the storm. I ducked under cover and we talked with each other while the rain pounded the tin roof over our heads. She told me she was an artist living in Pigeon Forge, Tennessee and enjoyed coming into the Smokies for frequent day hikes. She showed me on her cell phone some of her wildlife drawings and paintings, explaining that she had all her art on display in a gallery there. Later the rain stopped, and she hiked back to her car, parked about five miles away, at Clingman's Dome parking lot. I was pretty sure that she had no rain gear or even an umbrella. In less than 30 minutes, the heavy rains and thunderstorms returned, and I thought about the poor girl being drenched.

Later that night, I made tuna and beef stick rice for dinner. This was one of many odd combinations of common ingredients from my food bag that would only appeal to a thru-hiker: surf and turf, hiker style. After my gourmet feast, I went down to get water, hung my food bag, and got set up for the night in the shelter. Very heavy rain continued all night long; it was a good call to be staying under cover tonight for only the second time on my great adventure. Ironically, the first night I stayed in a shelter was at *Siler* Bald, now tonight it was *Silers* Bald. My feet were really hurting and I was considering where and when I might take another ZERO day.

Day 16
14.7 miles

Day 17 June 22, 2020 Monday (GSMNP Day Three)
Silers Bald, NC to Mt. Collins Shelter, NC

I stayed in Siler's Bald Shelter alone last night, and it rained pretty much non-stop until a few hours before sunrise. Shortly thereafter, I woke up to birds chirping at first light. For breakfast I made cold-soaked oatmeal and one cup of hot coffee. No bears or any other distractions other than a lot of wind and rain all night long. I hit the trail late about 7:45 AM, hiking along the Narrows with fog all around me most of the morning. The Narrows is a ridgeline part of the trail where one to two feet on either side would precipitously drop off for several hundred feet at a very steep angle. The 4.5-mile hike up to Clingmans Dome was a steady uphill with abundant terraces and small gaps along the way. I began seeing a lot more conglomeratic sandstone amongst the assortment of other metamorphics beneath my feet as I slowly gained elevation. It was a surprisingly easy hike relative to the previous day's trek from Spence to Derrick Knob over Rocky Top.

I met a lot of folks on their way up to Clingmans, including a young lady who was in a boulder-strewn meadow photographing the Dome and other nearby peaks. I would run into her later on my way out of the Smokies, past Davenport Gap. When I arrived at the overlook area for

Clingman's Dome, there were dozens of tourists, mostly coming from their cars in the nearby parking lot. Some of them were really interested in hearing about my thru-hike. It was fun fielding questions from the youngsters and even their parents about what it is like walking through the woods all day and sleeping on the forest floor each night. I stood around talking for quite a while to several groups and families, and after about 20 minutes, continued on my way north.

Clingmans Dome is 200 miles north of the southern terminus, and at 6,643 feet, is the highest point along the AT. There is no alpine zone in the Smokies, and although there are many grassy balds on the southern mountains along the AT, Clingmans is not one of them. Therefore, a large spiraling concrete ramp leading up to an observation tower was constructed in 1959, affording incredible 360° panoramic views of the surrounding Smoky Mountains and beyond. I have been to Clingmans at least three times before, each one in overcast or foggy conditions where the views were either cloaked in haze or non-existent. Today however, I would finally get a clear view of that panoramic spectacle!

After Clingmans, the hike was generally downhill to Collins Gap, just before Mt. Collins. What made this area so special was the dense spruce forest on the ridge top, extending all the way to the shelter area located on a nice flat less than one half mile north of the AT. Soon the rain returned, and I stopped to cover my pack and put on my trusty blue rain jacket. While trudging up the final ascent, I passed several day-hikers including two teenage boys who came running past me at full speed down a fairly steep rocky part of the trail. Had I been 50 years younger and minus this 50-pound pack, I might have been able to run perhaps half their pace.

When I arrived about 2 PM, I decided that I needed to rest my legs and feet for more than a few hours. I ducked under the cover to take off my pack and rest a bit when I heard someone slogging through the muddy puddles outside. I turned to see a young man named Josh sitting by the table beneath the overhang watching the rain. I came out and we talked for a while, and when I asked him where he was from, he said, "all over," but wouldn't be specific. Apparently, he was a musician and traveled around the country with various bands. He had set up his tent nearby and had been hiking for the past several weeks on many of the lesser known side trails throughout the park. He explained that he had previously completed a thru-hike of the AT, but that most of the trail follows ridgelines, or is in the higher elevations in general. He went on to say that there are so many hidden gems like waterfalls, low country bogs, and meadows that the AT misses and that was the reason for his present hiking trip.

I was originally planning to stay in the shelter once again, due to the rain and dropping temps. About 6 PM, two more hikers arrived, and it was Emily and Aiden from two nights ago. At first I did not recognize them in their rain garb in the dimming daylight, until their dad Brian and brother Ben arrived a few minutes later, along with Tigerfrog. It looked like we would all still fit inside, but then a lady and her seven-year-old nephew appeared. I decided to just move out and set up my tent on a nearby mossy flat in order to give everyone else more room. This proved to be a good call, because the young boy had left his sleeping bag behind, requiring him to share with his aunt for the night. He evidently didn't sleep well and I could hear his frequent cries as he continually woke up multiple times throughout the night.

Overall, it was a very easy and relaxing day, just what I needed to give my feet a break. Despite the thunderstorms most of the afternoon, the early evening brought with it the return of the sun, giving the day a very pleasant finish. However, and not surprisingly, the forecast was for rain all night.

Day 17
7.5 miles

Day 18 June 23, 2020 Tuesday (GSMNP Day Four)
Mt. Collins Shelter, NC to Peck's Corner Shelter, NC
I woke at 4:48 AM to the sound of Brian and his three kids packing up so that they could get home to Chicago today. There was the red glow of their morning headlamps flickering all around as they gathered up their gear for their hike out. Although it rained overnight, the morning was now calm and still. I had heard from Josh's weather report that early morning storms were likely, so I began packing up early and was on trail by 6:15 AM.

It was wonderful hiking in the quiet and mysterious early light of the predawn. (As the days grew shorter in the latter days of my journey, I would experience this daily). My headlamp illuminated the uneven trail directly in front of me, which I constantly focused on in an attempt to not trip and fall. Peripherally, I could sense the thick, fragrant darkness of the surrounding spruce forest traversed occasionally by raging cold mountain streams. Moss covered metasedimentary boulders littered the shadowy forest floor, causing me to peer occasionally into the gloominess for any sign of movement. I was keenly aware of the reported large bear population in the Smokies and calculated that these would be the ideal conditions for such an encounter. However, just like every other day so far, I saw no bears.

Just before I got to Newfound Gap I ran into Bonanza and his brother Rocky. I was elated that they decided to hike through the Smokies, and it was also a surprise to see the two of them ahead of their dad, Rocket-T1D-73. They explained that he wasn't feeling well today and so had lagged just a tad behind. During our discussion, I explained how I needed to soon resupply but wasn't sure where or when I could do so. I certainly did not want to take the detour into Gatlinburg for just a few items. In response to my plight, they graciously offered me some snack items to hold me over for another day, which I truly appreciated and sorely needed! We talked for a while longer, snapped a group selfie and then parted on our separate ways.

When I arrived at Newfound Gap, there were of course many tourists milling around the large parking lot, with many day-hikers setting out in both directions. Newfound Gap is a bit over the halfway mark for NOBO's and many hikers take this opportunity to hitch rides into Gatlinburg via scenic highway US 441 for a resupply, or to take a ZERO day. For me this was simply an opportunity to stop and rest my feet for a few minutes. A few curious tourists spotted me with my backpack and strolled over to ask questions and chat about my hike. After a few minutes of talking about my adventures with a young couple, they walked back to their car, then suddenly the man came running back over to offer me a fresh pack of Dentyne gum. I thanked him and thought, once again, what a kind gesture this was. I know this sounds crazy, but I never tasted such delicious chewing gum before in my entire life. Later, I ran into Rocket-T1D-73 but he was in a hurry to catch up with his two sons, so we just exchanged greetings and continued on.

South of the gap, the trail was rocky and very steep in some places, but it was generally a gradual uphill for several miles. After almost three weeks into this hike, my legs were getting acclimated to these long uphill climbs. Over the next several miles, I passed about a dozen southbound day-hikers on my way up the trail toward Icewater Shelter. I arrived there at 11:30 AM, had some lunch, repaired my toe and heel bandages, and rested for about 20 minutes. While still under cover, an intense rain started, lasting for about 30 minutes. Following the downpour, there was a change in wind and temperature, and it felt much colder afterwards.

Shortly after lunch, I came to Charlie's Bunion, named after one of the early mountain guides in the Smokies. The exposed bedrock here was primarily slate, part of the Anakeesta Formation, also an assemblage of metamorphosed sedimentary rocks of the Ocoee Supergroup, which was later thrusted westward along the Great Smoky Fault. It seemed to me unusual to have a rock, which first started out as shale, to be a ridge-forming unit. Most shales are easily eroded and weathered, forming valleys rather than ridges. There was an interesting sign at the entrance to the side trail, warning parents to "closely control children" on the trail. Here the trail passes along a treeless and precipitously steep rock exposure where one wrong step could lead to tragic results. (It is worth noting that although I did not get to meet her, I had heard several times about a 2020 thru-hiker named Anakeesta, who, like myself also had a trail name after a geologic formation.)

The views here were spectacular, amid the low clouds slowly breaking up to reveal the distant peaks. For the remainder of the day,

I hiked in the off-and-on rain along open ridgeline trails, with incredible views in both directions. This might have been the best day yet for breathtaking views. I ran into several other hikers along the way, notably one young couple who were both in shorts, flip flops, and tank tops, with no umbrellas or rain gear. They had to be freezing!

I finally arrived at Pecks Corner about 5 PM, which was 0.4 mile down a blue-blaze trail. There were four or five guys inside; most of them were chefs and restaurant workers from Indianapolis, who had just been laid off. They had a tarp across the front of the shelter with a blazing fire going inside in the fireplace. There are only a dozen or so that actually have inside fireplaces, and this was the first one that I had seen. One of the guys in the group had apparently burned one of his shoes trying to dry it out over the fire; it was rumored that this was his only pair. I never heard how, or even if, he got out the next morning. I also met Wade from Roswell, a real pleasant guy about my age. We had a lot in common and talked about our families, interests, and hobbies for most of our dinner hour.

I had earlier set up my tent in a little secluded spruce thicket, along a side trail to a ridgetop campsite overlooking the picturesque valley below. There was cell service at the overlook, but at nearly 5,500 feet elevation, it was so windy and cold I thought it prudent to stay lower in the protection of the dense tree cover. It had been a long day, and I retired to my tent by 9 PM.

Day 18
15.4 miles

**Day 19 June 24, 2020 Wednesday (GSMNP Day Five)
Peck's Corner Shelter, NC to Cosby Knob Shelter, NC**
I woke with the sound of the birds to a foggy, cool morning. My isobutane canister was spent, but fortunately I had brought along a spare despite the added weight. I used the new gas canister to boil water for my oatmeal and coffee, my usual breakfast of champions. I hit the 0.4-mile uphill blue-blaze trail about 7:30, and about an hour later, Geoffrey, a French hiker who stayed at our camp area last night, caught up with me. We hiked together for the next four or five miles chatting most of the time. It was nice having someone to converse with while I hiked, something I rarely experienced. He worked for a French energy company here in the US, and so we had a lot of things in common to discuss and debate. After about two hours, I could tell that his pace was a little faster than mine, and eventually he moved on.

Other than two women that I ran into that morning, I didn't see anyone else on the trail. Although it was very foggy, there were a lot of great overlooks along the way, but with hardly any views. Around noon, I stopped for lunch. My feet were still very wet and so I changed some bandages and tape, ate, and took a short 15-minute break. Nearly the entire day so far was at elevations over 6,000 feet, with a slow 1,000-foot descent beginning after lunch. Around 3 PM, I began a steady climb back up to Crosby Knob.

I got to Peck's Corner around 4:30 PM and met Crockett, from Philadelphia, Pennsylvania. He was a NOBO thru-hiker who made it as far north as Hiawassee in March, then flew up to New Jersey and started south. He told me about some of his adventures, such as having to hide behind trees to avoid being seen by park rangers while in the Shenandoahs. He also told of a bear encounter he had while in his tent one evening, and how he frightened the creature away by playing a Janis Joplin song at high-volume. We both tented near each other down a hill away from the main camp area. That night, a bear tried to get into the shelter, apparently attracted by the smell of food. He also climbed the trees from which the bear cables were strung to hang our food bags, but was unsuccessful at reaching any of them. We could hear all the men yelling at the poor animal periodically for several hours as the bear made his repeated attempts.

Tomorrow, I planned to hike eight miles down to the eastern boundary of the Smokies at Davenport Gap, cross the Pigeon River and I-40, and end at a campsite somewhere near Green Corner Road. Altogether, I would be losing about 3,500 feet elevation over roughly 11 miles.

Day 19
12.9 miles

Day 20 June 25, 2020 Thursday (GSMNP Day Six)
Cosby Knob Shelter, NC to Standing Bear Hostel on Green Corner Road, NC

I awoke to birds at 6 AM, packed up the tent, and brought all of my gear up to the shelter. All of the morning talk related to the multiple bear visits last night. The bear apparently chewed a couple of holes in Eric's plastic bowl, which he left on the table. Other than that, there was no damage. I had my coffee and oatmeal and set out for the trail at 7:45 AM.

It was a late start, and I was prepared for about seven or eight miles of all downhill. However, after Low Gap, I hiked steadily uphill for over an hour to the fire-tower trail. From there it was mostly downhill with occasional uphill climbs. On my way down I saw a lot of bear scat. I heard heavy stomps and grunts in the thickets, but still no actual sightings. Most of the bedrock here was Proterozoic age metasediments, predominantly Thunderhead Sandstone of the Great Smoky Group. These are metamorphosed massive layers of feldspathic sandstones and graywackes, which have been repeatedly folded and thrusted over geologic time. It had been misty and raining off-and-on all morning and by noon the sun eventually came out. I ran into a lot of day-hikers who were coming up from Davenport Gap to the Mt. Cammerer Fire Tower. The fire tower was on a blue-blaze side trail, and because of the overcast conditions, I decided to just continue down to Davenport Gap where I arrived about 1 PM. From there, I followed the AT blazes to Pigeon River, I-40, and up the stairs to the trail leading to Green Corner Road where Standing Bear Hostel is located.

I ran into one other backpacker who was heading in the same direction, and we walked together up the road to Standing Bear. I thought she looked familiar, and it turned out that we had met a couple of days earlier where she was taking pictures just before Clingmans Dome. She was from Tennessee and also needed a resupply and a break after the Smokies. When I arrived at the hostel around 3:30, it was more like a very cool and rustic little village. It reminded me of something right out of the 60s. There were a lot of beads, colored stones, and generally that hippie motif. There were a few hikers wandering about, all showered up and apparently well fed—at least recently. I met Jeff, who appeared to be in charge, and we sat on the porch and talked for a while. Jeff was a sincerely kind, attentive, and amicable young man, but as we talked, I couldn't help noticing his conspicuous lack of teeth, other than one on the top and one on the bottom. He told me he was a surfer, originally from California, who thought this would be a great place to come and work for a while, and that his girlfriend (who I did not meet) was a competitive skier. He then explained to me about the resupply store and then presented me with his recommendations on the pizza and beverage options.

I bought my food and supplies for the next several days, and then decided on the pepperoni pizza and ice-cold IPA option, while I waited for my phone to recharge. While feasting on this heavenly lunch, I chatted with the lady from Tennessee about the AT, but I think we talked mostly about our families. Later, I left Standing Bear and walked down the dirt road, crossed over the bridge, and found a nice tent-site right next to the stream. I set it up, then took a refreshing dip in the creek to wash off, and finally boiled some water for my noodles. I had dinner and was in my tent before dark.

Soon after—and for the next several hours—there was an inordinate amount of traffic coming up that old dirt road, apparently heading for some type of event going on at Standing Bear. I thought about walking up to see what was happening but was seriously trying to get some sleep after six long days in the Smokies. Later in the evening, all that traffic returned back down the road, headlights lighting up my campsite as they passed by.

Day 20
10.7 miles

Chapter Four
The Balds

Day 21 June 26, 2020 Friday
Standing Bear Hostel on Green Corner Road, NC to Max Patch Mt., NC

    I awoke at 5:30 AM, re-bandaged my toes, broke camp, had breakfast, and was on trail by 7 AM. This morning was a very long ascent up Snowbird Mountain, beginning at 1,800 feet and gaining over 2,450 feet in elevation to the summit, at 4,263 feet. There were several nice campgrounds that I passed along the way, which caused me to think that maybe I should have gone a little bit farther yesterday. The hike up the mountain was not too difficult, the grade was very reasonable, but the distance was nearly five miles.

    When I finally arrived at the top of Snowbird Mountain, just before noon, there was an FAA tower on the bald top looking like something from an old James Bond movie. There were great views in several directions, and I leaned my pack against some Longarm Quartzite boulders to rest and have a morning snack. These were more of the metasediments similar in age and origin to those I had been stomping my way through in the Smokies for the past week. I made it a point this morning to stop several times and snack on the way up in order to keep my energy level high. When I did stop for lunch, I sat down, leaning against my pack with my shoes off for about an hour. This seemed to really help for the afternoons.

    By about 3 PM as I was approaching Max Patch Road, I met Journey Man, a 2016 NOBO thru-hiker who had a beautiful hand carved hiking staff. He told me that there was some "trail magic" ahead, and that someone was giving out hot dogs, snacks, and cold drinks to all the thru-hikers who came by. In less than a mile, there he was—a kind gentleman camped alongside the road, and when I walked by, he called me over to offer me a beer or a cold drink. I sat down with him, and we talked for about 15 minutes while I finished my refreshing beverage. He worked in HR for a Japanese petrochemical company in Spartanburg, South Carolina. Before leaving he pointed me in the direction of a cold running mountain spring, where I could refill my water bottles. I thanked him and was on my way.

    While at the spring, I considered going past Max Patch and onto the next shelter just to put in more miles for the day, but so many of the hikers I met had told me not to miss camping out on the expansive grassy bald there. Until I got to this point, I really wasn't sure what I wanted to do. After thinking more about it, I decided to stay on top of the mountain tonight, and just hike those extra couple of miles tomorrow. It was a long but interesting hike to the treeless top of the mountain, and there were a lot of other people who were also heading in this direction.

    When I arrived at the top, it looked like a large open field, with tents pitched all around. Occasional boulders of Max Patch Granite were present near the trail, but overall, the meadow-covered bald mountaintop kept its secret interior well hidden. The whole place had a festive feel to it. Young people were throwing Frisbees, dogs were running free, and some folks were simply sitting in the grass while gazing off into the horizon. It reminded me of spring break at the beach. I walked around until I found a spot to claim as my campsite, and I unloaded my pack and set up my tent. It was 4:30. I walked around for about an hour just taking in the sights and the scenery.

As I was staring off toward the west and the Smokies, I noticed someone slowly trudging up the path towards me, wearing two metal knee braces. It was Eric from several of the past camp night stays, including the night when the bear chewed up his plastic bowl. I called over to say I had a camping spot, and that he was welcome to join me. We both walked back to the campsite, and after he set up his tent, we sat down together and made our dinners. Next to us, there was a group of four people sitting around their campfire, including a young lady with a puppy that kept running in and out of one of their tents. She and her dog came over and asked if we were thru-hikers, sparking a lively discussion about all of the adventures I had experienced on the trail so far.

Later I walked back over to watch the sunset and noticed there were dark storm clouds obscuring my view of the sunset. Evidently those clouds were pouring down rain in the Smokies and heading our way. It took about another half an hour until the rain reached us, and I made it to my tent just in time. It turned out to be just a passing shower, and within the hour it was clear again. That night, I mostly just listened to the wind and the sound of young people all around me talking and laughing.

Day 21
13.6 miles

Day 22 June 27, 2020 Saturday
Max Patch Mt., NC to Deer Park Mt. Shelter, NC

Last night, I eventually put in my earplugs to drown out the partiers. At 5 AM, I woke up to my tent violently shaking with wind. I had a bad night sleep on the mountaintop, repeatedly waking up in pools of sweat. My current sleeping bag was an L.L.Bean goose down designed for temps as low as 20°. It was great for those cold nights like I had in the Smokies at 6,000 feet, but for most nights this summer, it was far too warm. For the past several nights it was a constant covering and uncovering of the top of my much-too-warm bag, never able to hit that optimum temperature for restful slumber. If I wanted to sleep well in this summer heat, I would soon need a new, more lightweight bag.

When I took out my earplugs, I heard the wind whipping through the tent. I also heard lots of laughing and shrieking girls whose tents were being blown away or knocked over. Then came the rain. I got dressed in my tent and hurriedly packed up my sleeping bag and air mattress. I then loaded up my backpack under the cover of my vestibule, tore down my tent, and by 6:30 AM, was hiking down the mountain in the windstorm. Along the way, I saw lots of tents blown down and people running about packing up to get off the mountain, too. All the effort put into drying my socks and shoes yesterday went to waste within the first 30 seconds of walking through the soaking wet grass.

It turned out to be an enjoyable hike down the mountain, and by 8 AM, I was at Roaring Fork Shelter to fill my water bottles. However, I was still feeling like I made a big mistake camping at Max patch, having lost time and adding those extra miles to an already-planned long day. Two hours later, still feeling behind the eight ball and with sore feet, I stopped at a clearing to mix up some cold coffee, accompanied by three ibuprofen and a Clif bar. I also took off my shoes to inspect my feet and add a few more bandages, and for the rest of the day, I was more energized and had a less painful foot. I tried to stop every two hours and have another bar as they seemed to really help my energy level. It was becoming clear that I needed to eat more during the day because I was burning far more calories than I was consuming.

I finally arrived at Deer Park Mountain around 4 PM. I decided to camp out here tonight rather than going all the way to Hot Springs, which was another 3.2 miles. It turned out to be a very good call. At the trail juncture, the shelter trail went to the right, and the water-source trail went left. Two logs were lying across the blue-blaze trail to the water source. Beneath the first log was a fair-sized timber rattler just curled up and minding his own business. I looked behind the log and someone had evidently set up a tent right in the middle of the trail. Looking up the hill a little farther, I noticed there were two people lying on a blanket together, and I called up to them "Is this your tent?" The young man looked up, probably wondering why anyone was bothering him, and I continued, "It looks like there's a timber rattler right next to your tent." The girl immediately jumped to her feet and began shrieking, "I knew it! I had a dream last night that there were little worms all around me and they all suddenly turned into snakes!" She came running down from the hill in a panic and was very upset.

I then headed down the blue blaze trail towards the shelter and found a nice tent spot. After setting up my Altaplex and unpacking gear, I walked over to meet several other hikers, including a scout leader, Terry, who was backpacking for a couple of days with three of his boy scouts. Another hiker there, EZ-E, worked on Blackhawk helicopters. We all had dinner together and talked for maybe another hour until I turned in for the evening. Other than the brief morning cloudburst, it had been a beautiful day.

Day 22
16.6 miles

Day 23 June 28, 2020 Sunday NERO day
Deer Park Mt. Shelter, NC to Laughing Heart Hostel, Hot Springs, NC

I awoke early this morning to a steady rain at Deer Park Mountain camp. I lingered in my tent until the rain subsided about 5:30 AM, then got dressed and broke down camp. I went over to the shelter, had breakfast with other hikers, and was first to leave. This was intended to be a NERO day, and I wanted to arrive at my hostel as early as possible. In the hiker world, a NERO day simply means "near ZERO," and suits me better than taking a full day and two nights off the trail. The initial uphill climb caused me to shed my rain jacket. About a half hour later the rain resumed and back on went the jacket.

By about 9:30, I completed my 3.2 miles for the day, arriving at the edge of the woods, and entering the resort town of Hot Springs, North Carolina. Two young hikers who I continually passed as they took their many breaks, and who would then leapfrog past me later, came running by me one last time within 50 yards of the forest edge, as if they were waiting for that moment all morning. It reminded me of the tortoise and the hare, but this time the hares won! Hot Springs is named for—as you might have guessed—its famous hot springs. The reason the springs are there has to do, of course, with the geology. Discovered in the late 1700's, the thermal springs are the result of groundwater travelling down a fault to where it comes in contact with much warmer rock, then circulating back to the surface retaining much of the heat from below.

Right there at the end of the NFS parking lot, and directly adjacent to the AT, was the Laughing Heart Hostel. There were several people standing out front talking, as if they were waiting for me to arrive. I walked over to say hello and introduce myself to Tigger and Chuck Norris, proprietors, a very kind and warmhearted couple. After checking in, I was assigned a private room right next to the kitchen. I proceeded to unpack all of my wet gear, hanging things all around the room to dry. I took a long steamy shower, did laundry, and found a comfy chair on the back porch to sit and make some calls and texts.

After about an hour of relaxing on the back porch, I decided to walk to town for lunch and resupply. It only took me about 10 minutes to arrive at the town center, but I foolishly wore my sandals with no socks, which didn't feel so great on my already sore feet. I spotted a busy barbecue joint on Main Street, walked in, and ordered a scrumptious barbecue brisket sandwich and a tall, ice-cold lemonade. After lunch I walked down the street to a tavern someone had recommended. They were advertising a large ribeye steak, which sounded just right for dinner later that night.

On my way back, I stopped by the Hillbilly Market for my resupply and returned around 4 PM. Back at my room, I re-organized and packed my food bag, rested, and made a few more calls and texts. I noticed a scale behind the laundry area for weighing packs and other smaller items. I went outside to weigh my food bag, and the needle on the scale registered almost 15 pounds. Wow! Too much! Next I returned to the patio area, sat down on a comfy chair and made airline reservations for my return trip to Pittsburgh on July 14.

Around dinnertime, I returned downtown along with my appetite ready to attack that ribeye at the Iron Horse Tavern. It truly was as wonderful as I had imagined. After my feast, I returned to the hostel, but following the white blazes of the AT this time. The route led me up a long narrow set of rock steps to emerge from the woods right behind the hostel. I sat outside for a while chewing the fat with Chuck Norris and a few other hikers while the sky began its nightly pastel transformation. My plan was to leave early in the morning and so I turned in early. It was nice to sleep in a real bed for a change.

Day 23
3.2 miles NERO day

Day 24 June 29, 2020 Monday
Laughing Heart Hostel, Hot Springs, NC to Allen Gap, NC

I woke up suddenly to the drone of a steady rain outside my window, surprised to be lying in a real bed at Laughing Heart Hostel. I came out of my room and walked across to the kitchen area and found oatmeal and some hot coffee for breakfast. I decided to wait out the rain, leaving about 8 AM for the trail. Several other hikers had already left in the heavy morning rain, for which I didn't see the point.

I walked back down the rock steps to the street and then back through town following the AT across the French Broad River. I was feeling good about my decision to wait out the deluge, and I could already see patches of blue sky through the fog and clouds. After the bridge, the trail left the street, and I followed the river for about a half a mile, slowly gaining elevation. Somehow, I missed the first switch back and continued on an unmarked trail thinking it was the AT. The trail suddenly began a very steep ascent up the side of the mountain for 50 yards or so, before I decided this just couldn't be right. I backtracked, quickly saw my mistake, and got back on the AT.

The trail was very narrow, but nicely graded, taking me on a series of switchbacks cut into the steeply dipping Cambrian quartzite bedrock, all the way to the top to Lovers Leap. From there I could see the French Broad River and Hot Springs below, with cloud-shrouded mountains in the distance. According to Native American lore, this was the spot where a young Cherokee maiden, Mist-on-the-Mountain, allegedly plunged to her death after learning of the brutal murder of her lover, Magwa, by another jealous suitor. As poetic justice would have it, the story continues that her jealous suitor, a young brave by the name of Tall Pine, was later attacked and killed by a wild panther.

I made stops for energy bars and water about every two hours, also taking two ibuprofens at my first stop. This schedule of breaks seemed to be working well. There was an occasional sprinkling of rain but not enough to even require a rain jacket. The weather improved over the course of the day becoming warmer and sunnier.

Somewhere before Pump Gap, I heard voices and footsteps approaching me from behind, and two trail runners passed by. Shortly after, I met Sparky from Tampa, Florida who was just out for a 30 miler, and we hiked together for most of the day. Later, we ran into the Honeymoon Couple, who apparently were married on the trail, and have hiked it four times. I was going to finish the day at Spring Mountain Shelter but, when Sparky and I arrived, I decided I needed to put in more miles for the day. Sparky stayed but I went another 3.7 miles to a campsite along the trail just a stone's throw from Allen Gap. This was one of my first stealth sites where I would camp near the trail on a small flat, often just large enough for one small tent. Occasionally I would select one of these sites after long enduring hours, where I could walk no further. Today my persistence paid off, yielding to a cozy secluded spot amongst the thick cover of lush rhododendron and hemlock.

Day 24
14.7 miles

Day 25 June 30, 2020 Tuesday
Allen Gap, NC to Snake Den Ridgecrest, NC

I woke to a Barred Owl hooting somewhere nearby, and eventually I began hearing birds about 5:15. I broke camp early in anticipation of possible rain and was on trail by 6:45. I hiked to the next shelter for my second breakfast and privy, just a couple of miles from my campsite. I continued following the trail, next to a swift babbling stream, when I came upon a massive green wall of moss and ferns completely blocking the narrow valley in front of me. Its unexpected presence caused me to do a double take, and upon further inspection, I realized that what I was facing was a very old, massive concrete dam. I followed the trail to the top where I found a surprisingly clear and serene pond behind it, full of hatching flies, frogs, and fish. There were several well-worn gravel patches along the shoreline where fisherman had often stood, perhaps for hours, in their attempts to land a lunker.

I hiked past the pond where the white blazes led me through fields of high grasses through which I later emerged onto an old country road. I followed the road for about a half-mile and eventually was guided back into the woods. I began climbing upward on a steeper section of trail toward Camp Creek Bald, reaching higher elevations where the trail passed over many flowing springs. There were several boulder scrambles along the way, making the trail much more interesting. These rocks had changed again, appearing more rugged and taking me over 1,000 feet higher than the older Precambrian peaks I summited yesterday. I could see that I was now back into the early Paleozoic quartzites.

While hiking amongst the myriad of springs, I ran into Rocket-T1D-73 and his two sons Bonanza and Rocky, each hiking alone about 10 minutes apart. I stopped to talk to each one of them, over the course of the next 30 minutes or so. They all were excited to tell me about the trail ahead, and how it followed an exposed ridgeline with occasional glimpses of breathtaking views when there was a break in the fog. I continued toward White Rock Cliffs and Big Firescald Knob. Over the next two or three miles there were numerous rock steps and boulder scrambles over these Cambrian-age metasediments. These were the same rocks I had studied for many years as a geologist in search of hydrocarbons in middle Cambrian sandstones and shales in West Virginia and Kentucky. Rather than lying beneath 8,000–13,000 feet of lithified sedimentary section in what geologists refer to as the Rome Trough, these rocks were here at the surface forming the ridgeline upon which I was now walking. As promised, the trail led me along a high ridge walk with lots of great views through the light fog to the east. Looking west, there was nothing to see but a thick blanket of blurred white. Up ahead, I came across Howard's Rock.

Howard's Rock was the highest point along the ridge trail, and I stopped there to read his one-page commemoration, which was laminated in clear plastic and attached to the rock. Evidently Howard McDonald was a long-time trail worker and spent years helping develop this part of the trail. The rock bearing his name was one that he always thought should be moved to a higher point of prominence, where hikers could sit and enjoy the amazing view. So, one day, six fellow trail workers (likely stronger and younger) hoisted the rock to that prominent position as a surprise for Howard, where it remains today.

Parts of the ridge walk were quite treacherous and eventually the trail led me down the rocky ridgeline to lower elevations where I finally arrived at Jerry Cabin Shelter. A father and son from Hot Springs were just about ready to leave for Big Butt Mountain and on to Flint Shelter for the night. We talked for a few minutes, and as they were leaving, I went down to the spring for water in order to make it the 6.4 miles to my planned destination for the night.

Next, I climbed Big Butt Mountain, which was very steep and eventually opened to a large alpine meadow. For the next half mile, I walked on a narrow grassy path surrounded by a dense thicket of blackberry bushes stretching as far as the eye could see. The route to the top of the mountain was again a slow scramble over endless large quartzite boulders ultimately leading back to another alpine meadow and a pleasant downhill trail for several miles. I was thinking about stopping for the night before too long, but frankly, I wanted to get past the Shelton grave first. The story was that during the civil war, two locals joined the union army, and upon returning home for a family gathering, were ambushed and murdered by confederate soldiers; somewhere close to where I was now walking, alone.

The trail now was paralleling the ridgeline about 100 yards to one side. When I was about two miles from the Flint, I spotted a side trail leading uphill, turned, and followed it to a spacious ridge top campsite. The air was thick with little black gnats. I quickly set up my tent and unpacked for the night, careful to not leave the screen door unzipped any longer than necessary. Dinner was spam with pork ramen noodles. I hung my food bag then turned in to review maps and plan for the next day.

Date 25
14.6 miles

Day 26 July 1, 2020 Wednesday
Snake Den Ridgecrest, NC to Grassy Knoll 0.6 mi. before Street Gap, NC

I awoke with first light, heard owls hooting, and began preparing for the new day. Following my usual breakfast of oatmeal and coffee, I broke camp and was on trail by 6:52 AM. Even at this hour, I could feel the heat and humidity beginning to return. It was now into July, and the weather accordingly was becoming warmer and more humid by the day. Even on days starting out clear and sunny, the big cumulus clouds would begin to move in after lunch, bringing at least an hour of rain almost every day. The first thing I would hear would be far off thunder, and then the winds would pick up speed, and I would hear it in the trees, slowly increasing its intensity. Soon, there would be the sound of raindrops pelting the leaves like a faraway freight train slowly approaching. I would have about 30 seconds to get prepared. First, I would put my cell phone in its plastic waterproof pouch, then pull out my pack cover and cover my pack. If I were hiking on a flat or a downhill, I would quickly put on my rain jacket. If I were on a long climb, I would generally let the rain soak me while I continued sweating my way uphill.

From my stealth campsite, I returned to the trail, where I was met with a smooth downhill and easy gradient for the first mile or so, then a steep and precipitous section down to Flint Gap. After a short climb up and over a small knob, it was generally downhill on good trail to Devils Fork Gap. Here I crossed several roads and pleasant pastoral areas. After the gap, the trail became very rocky, feeling very much like Pennsylvania!

Around noon I finished the very long steep ascent to Sugarloaf Gap and along the way, I met the father and son from Hot Springs again. When we arrived at Sugarloaf Gap, we all stopped and had lunch together in a clearing where we all sat leaning against enormous fallen logs. I recall the dad being very interested in hearing about my thru-hike plans, and I gave him some advice on some of the gear I liked the best. We also talked about our favorite energy bars, and the boy and I traded some.

After our break, I continued my ascent up to Lick Rock, and although it was not as steep, it was still a long and difficult climb. Rain started and this compounded the difficulty of the climb. I continued past Big Flat Campground and finally up and over High Rock, after which the trail maintained a nice downhill grade all the way to Sam's Gap.

When I emerged from the trees at Sam's Gap, I followed the path down a gravel road, which led me to the I-26 underpass. It was still raining, and as I walked under the bridge, I met a thru-hiker, Honey Mustard, and two of his friends with their dogs. All three of them were chefs from Boulder, CO, and unfortunately unemployed due to COVID. The two friends were visiting Honey Mustard for a week or so while he hiked north into Virginia. We all talked under the cover of the bridge until the rain stopped.

The restaurant business had been decimated in 2020 because of the COVID pandemic. People were no longer going out to bars and restaurants, and many of the laid-off workers were now out here on the AT. During my five months hiking I would meet dozens of chefs, bartenders, and restaurant workers who had lost their jobs.

Later as I continued up the trail, the rain started up again. Because I was in a fast, hard, uphill climb, I got soaked but it felt great and refreshing. The trail brought me to the summit of the next mountain where I found several roomy campsites. Ordinarily I would have been delighted to camp here, but up ahead, what appeared to be a sunny clearing caught my eye. I walked a bit farther down the trail and found a nice campsite on a small bald along the ridgeline, just before Street Gap. From here, I could see Bald Mountain, my destination for tomorrow morning. To the east I could see a resort community on the next ridge over. It was 5:30 when I set up my tent for the night, made dinner, and while reclining against some indefinite metamorphic rock, watched an amazing golden-red sunset. I retired to my tent shortly thereafter, reviewing maps and planning for my day tomorrow. Later, I could see the moon high in the eastern sky shining like a spotlight through the wall of my tent. I had picked the perfect spot to camp on this night.

Day 26
15.2 miles

Day 27 July 2, 2020 Thursday
Grassy Knoll 0.6 mi. before Street Gap, NC to
No Business Knob Shelter, TN

I woke up about 5:30 AM with a pink sky in the east foretelling of a glorious sunrise. This was an incredibly scenic spot, but because of rain the previous day and the fact that I was set up in grass, it was very wet. I had hot oatmeal and coffee for breakfast, broke camp, and was on trail by 6:52. It was a nice early hike with sunlight filtering through the rising fog all around. The trail was fine until I was within 1.5 miles of Big Bald Mountain where the grade became very steep with many rocks and boulder scrambles weathered out from the surrounding biotite-gneiss bedrock. When I reached the summit, there was a beautiful 360° panorama. It was clear with just a slight haze to the north and east. I was able to spot my campsite from last night, seeming much too far away for me to have hiked all the way to this point in just a few hours.

Heading north, my maps indicated that the trail was nearly all downhill to the highway at Spivey Gap. Wrong! There were two other significant mountains to climb in between. More precisely, there were two very deep gaps ahead, out of which I would have to climb. Generally, the trail was very reasonably built, with good downhill grade, but the uphill climbs were insanely steep. The uphills out of the gaps to Little Bald and High Rocks were very strenuous and hard to climb.

When I arrived at the bottom of Spivey Gap, I thought the hardest part of the day was behind me. What I discovered was the initial climb to my next target of Devils Gap was ridiculously steep, following waterfalls and gorges straight up the mountain. This was still North Carolina, and later, for example in parts of Maine and New Hampshire, I would chuckle at my early ignorance. But for now, this was pretty extreme to me.

Once I got past the clambering up the side of the mountain, there were 3.2 miles of beauty and extreme quiet as the chiseled trail followed contours, leading eventually uphill again to No Business Shelter. I arrived at 6 PM, no one else was there, so I camped solitary for the night. There was also no water, no privy, and no bear cables. My dinner was noodles cooked with only half the regular amount of water, and I had practically none left to drink. Later, I looked through the logbook and the last entry was from someone by the name of Bells—just last night. Her comments mentioned the lack of amenities as well as the abundance of hoot owls. She proved right on both counts, and I enjoyed hearing the occasional hoots throughout the night.

Day 27
16.4 miles

Day 28 June 3, 2020 Friday
No Business Knob Shelter, TN to Beauty Spot, NC
Awoke to owls hooting before first light. Later birds began their morning singing. Because there was no water, I broke camp with a just box of raisins and granola for breakfast. The entire way to Nolichucky River valley I looked for water, but only found one small spring where I was able to fill my bottle to about 1/3 liter. That's all I had for the remainder of the 5.7-mile hike to the valley. On the upside, there were some impressive views of the Nolichucky River along the way as the trail slowly descended down a series of switchbacks to the river road.

I had planned to call a shuttle for a resupply when I got to the bottom, but when I arrived at the Nolichucky River Bridge, I looked across the street and saw a sign for "Uncle Johnny's." I walked over and found a large, covered deck and patio where several hikers were milling around. I went inside to look around the store, and to my pleasant surprise, found everything I needed for my resupply. I was also able to buy several delectable White Castle sliders, which I heartily consumed while recharging my phone and resting my legs. Later, while enjoying an ice cream treat, I spotted Honey Mustard and three of his friends waiting for their raft trip shuttle. They kindly offered me an ice-cold Coors light, and we sat and talked for a few minutes longer.

By 12:30, I was back on the trail with my destination set for Beauty Spot, another grassy bald atop a Precambrian granite mountain with breathtaking views all around. It was also more than 10 miles away, and all uphill. The first mile was easy, as the trail just followed a cool mountain stream with lots of inviting pools. It was a swelteringly hot afternoon, and I was actually thinking about taking a dip when I ran into my new friend Rocket-T1D-73. I told him about all the goodies at Uncle Johnny's. He seemed excited about stopping there and asked me to relay his plans to Rocky and Bonanza when I saw them. Ten minutes later, I happily informed them about ice cream and cold drinks in their future.

From there, it was a very long and tedious climb, but by 6:30 I finally arrived at the top of Beauty Spot. It reminded me very much of Max Patch, but on a smaller scale, and there were several other thru- and section-hikers camping there as well. I found a nice spot in some low grass not far from the others and set up my tent and unpacked things. Later, I walked over to where several hikers had invited me over to their campfire, and we sat around watching the sunset and chatting for about an hour. I met Sorrel, a young lady who sings in a band in New York City, and Dana and his Australian girlfriend, Jenny. Sorrel was apparently an avid forager of wild greens. Dana and Jenny's tongue-in-cheek trail names were Beauty and the Beast, respectively.

This had been one of my more difficult days so far, although I know I've overused that proclamation. However, the good news was that in order to make it to Damascus by my target date of Sunday, July 12, I only needed to hike an average of 12.8 miles per day. My plans were to fly back to Pittsburgh on July 13, the following day, for a ten-day break. When I later turned in for the night, the near full moon was back, shining like a spotlight, and lighting up my tent. In the distance, I could hear the faint sound of fireworks going off all around me on Beauty Spot tonight. Tomorrow was the Fourth of July.

Day 28
16.7 miles

Day 29 July 4, 2020 Saturday
Beauty Spot, NC to Clyde Smith Shelter, TN

I woke up to an amazingly bright pink pre-dawn at about 5:30 AM. I emerged from my tent about 6 AM to see a line of hikers standing quietly facing me. Apparently, I was between them and the spectacle in the eastern sky. I walked over to join them, and we all stood in a line quietly whispering things to each other every now and then, but mostly just watching as the sky continued to change from a pink to an orange, and then finally, the bright-orange speck of the sun peeked over the mountains. We stood and watched until the sun had come all the way over the horizon.

I returned to my tent, broke down camp, ate breakfast, and was on the trail by 7:30. The grass was very wet as was most of my gear, and after about the first 10 or 20 steps, my feet were soaking wet too. It was a long day and seemed to drag on and on. I took a 45-minute lunch break, starting off the afternoon hike at about 2:15. Unaka Mountain was by far the highlight of the day. Nearing the summit, I entered a beautiful mountaintop spruce forest, with everything covered in thick green moss and young spruce trees sprouting everywhere. This was a brief preview of what every day in Maine would be like.

I arrived at Clyde Smith Shelter about 5:30 after a very long and tiring afternoon. My feet were still bothering me, and it was difficult to walk without noticing them every step of the way. I had been trying to spot some Chicken of the Woods mushrooms for the last few days. I thought I found a fresh growth, and I stopped and cut some to take back to tonight's camp. When I arrived, Beauty and the Beast were there along with Sorrel. Bells came a little while later and I showed everyone the Chicken of the Woods and asked Sorrel if she thought this was the correct mushroom. None of us were certain, and since we didn't have 100% agreement, I tossed them aside and we didn't even try them. I would learn later that the mushroom I cut was not in fact an edible variety. From that point forward, I pretty much gave up on experimenting with fungi!

Another man, Matt, whose trail name was Mongoose, was camping nearby with his dog, Boomer. I asked him how he got his name, and he told me that several years ago he and his friends were out hiking, when one of them noticed a rattlesnake up ahead. Apparently, Matt walked over and nonchalantly raised and then proceeded to plant the spike of his trekking pole right through the head of the coiled rattlesnake earning him the distinction, from that day forward, of being known to all the world as Mongoose.

That night, it was only Sorrel and Bells who slept inside, and the rest of us set up our tents on some of the many flat sites around back. We all turned in around dark, and soon all I could hear was the quiet snores of other sleeping hikers, and the occasional voices and soft laughter coming from the shelter.

Day 29
16.8 miles

Day 30 July 5, 2020 Sunday
Clyde Smith Shelter, TN to Stan Murray Shelter, NC

I woke up this morning to the sound of gusting wind, and while expecting rain, came out of my tent to a clear and sunny morning; it was a nice surprise. I hung around the camp area with Dana, Jenny, and the others making breakfast and talking for a while. Today and tomorrow would be spent hiking the Roan Highlands, and the weather seemed to be perfect.

I was on the trail by 8 AM, with my first climb up and over Little Rock Knob, then down two miles to Hughes Gap to begin a grueling 2000-foot ascent in three miles to Beartown Mountain and finally Roan High Knob. The climb to Beartown Mountain was incredibly steep and rock-strewn and I thought I would never get to the top. Surprisingly, the next climb from Ash Gap to Roan High Knob summit was much easier, mostly due to a better grade on the trail. As I approached the higher elevations, the trail wove its way through dense spruce forests and over large mossy boulders of granite and Cranberry Gneiss.

Roan High Knob, elevation 6,285 feet, is the highest peak along the Roan Highlands. The hike up to Roan was gently sloping and at the top pretty much paralleled the ridgeline. Several hundred yards down slope to the northwest there were a few places where the trail led to minor balds along the ridgeline. However, most of these were treed in on all sides, not affording any views. Later I stopped at Roan High Knob Shelter for lunch and a short rest next to a large tree. Roan High Knob is the highest shelter along the entire AT, at 6,275 feet. After sitting down for a few minutes, I heard someone coming up the path, and it was Dana. He came over to join me for a snack, saying the others were about an hour behind. We talked for another 10 or 15 minutes, and then I was on my way again.

From my lunch stop, it was mostly downhill to the highway and parking lot at Carvers Gap. When I arrived, there were lots of tourists and day-hikers, with several people crowded around a large map next to the parking lot. I talked to several folks while looking at the map, most of them just out for a few hours on a nice, summer holiday weekend. I followed the AT to the top of Round Bald, Janie's Bald, and then on to Grassy Ridge Bald. Strewn about each of these balds were occasional outcrops of massive granites and assorted gneisses. The trail here was so much different than most of the AT, because even down in the gaps the ground was blanketed in grassy fields with very few trees around. It was a clear day, so I had amazing views the entire way from Carver's Gap to Grassy Ridge.

I continued north following white blazes toward Stan Murray Shelter for my night stay. At one of the springs, I noticed a brand new, large water bag someone had left behind. I thought about picking it up in the event I ran into its owner up ahead on the trail, but I thought it best to leave it since I didn't know the direction its owner had traveled. Further down the trail, I ran into Scott, and we hiked to camp together, arriving about 6 PM. He was there with his two grown children and his cousin Steve and two of his children. I spent dinnertime with his family. They were very friendly and easy to talk to, and we had a nice time together. They introduced me to canned smoked oysters, a new treat for me, which I would frequently stock during future resupplies.

Scott graciously offered to have me out to his Harpers Ferry mountain house when I came through later this summer. I agreed and said I would stay in touch as I got close to that point. I also asked if anyone in the group had lost a water bag several miles back, and unfortunately the answer was "Yes."

Most of the six other hikers had nice hammocks with rain tarps. I liked the idea of using a hammock versus a tent but for now would be sticking with my trusty 14-ounce Altaplex tent. My plan for the morrow was to re-supply for the rest of this leg of my trip at Mountain Harbor Hostel general store, 10 miles north of here at about 2 PM. At the pace I was going, I only needed to average 12.2 miles per day to reach Damascus by July 12.

Day 30
13.7 miles

Day 31 July 6, 2020 Monday
Stan Murray Shelter, NC to Mountain Harbor Hostel, US 19E, TN

This morning I woke to the sound of birds chirping and emerged from my tent to discover scattered rays of an orange sun breaking through the wooded eastern horizon. I got my gear packed up, had my hot oatmeal and coffee, broke camp, and was on trail by 7 AM. I was excited about today because I would be hiking over Big Hump Mountain. Back in early June, a fellow geologist and friend of mine told me Big Hump Mountain was one of the more memorable overnight hikes that he had taken when he was younger. Brandon was originally going to hike the AT with me but had some back issues preventing him from doing it. He sent me a picture taken back in 1974 of him and his longhaired friends sitting around a campfire at the summit. I had this on my list of places I was looking forward to since the start of my trip.

The other six hikers were still in their tents or hammocks when I left. It was a pretty good climb through the woods for about an hour, and then eventually the trail opened up to fields on top of Little Hump Mountain. I made it to the open summit by about 9:30 and there was magnificent scenery in every direction. It reminded me of the final scene from the Sound of Music. Ironically, while I was taking a 10-minute break, a father followed by six children emerged from the woods into the clearing not far from where I was resting. I did a double take, said hello, and then commented how much they reminded me of the von Trapp family!

When I returned to the trail, I was able to look across the bald ridgeline and see what was ahead of me for the next several miles. The views were truly indescribable. It was a clear day with just a few cumulus clouds dotting the sky above the far-off peaks all around me. After another hour of this ridge walk, I ultimately found myself atop Hump Mountain. All along the ridgeline today I was trampling over Precambrian Beech Granite or Cranberry Gneiss, with exposed ledges and boulders scattered here and there. I trekked through miles of grasses gently waiving in the warm summer wind, and behind that, blue tinted far-away mountaintops slowly fading with each successively more distant range until the mountains themselves were indistinguishable from the blue haze of the sky above. I located a nice flat spot near the trail to rest, reflect, and have lunch and promptly made myself comfortable. I took off my shoes and socks and relaxed against my backpack. For the next 10 minutes, I rested with my eyes closed with the sun shining down and a gentle breeze wafting across my face.

After a satisfying lunch and much needed rest, I continued on the blazed trail toward lower elevations. It was about five miles of mostly downhill to the hostel on Route 19. On my way down, I arrived at Doll Flats and spied the sign telling me I was leaving North Carolina. My cell phone battery was at one percent, and I was hoping to save enough juice for one last picture of the state-line sign. As soon as I took the picture my cell phone died! Just in time! This would be the first of many miraculous battery life extensions I would experience over the next several months.

At about 3 PM, I finally got to the bottom, then hiked up Route 19W about 0.10 mile to the hiker's entrance for Mountain Harbor Hostel. I checked in, had an ice-cold root beer, bought my supplies for the week, charged my phone, and then met some other hikers inside the air-conditioned common area. Later, Dana and Jenny arrived for the evening, followed shortly by Sorrel and Bells. We all shuttled down the highway to a Mexican place where we picked up our dinner. We then returned to Mountain Harbor and assembled outside, around several picnic tables. We talked over dinner about a variety of things including fossil fuels and alternative energy. After today, I only need to average 12.5 miles per day to make Damascus by July 12.

Day 31
11.1 miles

Chapter Five
Tennessee and the Road to Damascus

Day 32 July 7, 2020 Tuesday
Mountain Harbor Hostel, US 19E, TN to Moreland Gap Shelter, TN
 I woke up very late because my earplugs were in due to Route 19 traffic noise. I quickly dressed, packed up my gear, arose from my tent, and ran over to the bunkhouse. I went inside to the kitchen area had some coffee, picked up my charger, and left around 7:30. I was on trail down US Route 19E by 7:50 AM. The trail began in the woods but soon became a narrow footpath through grassy fields. It was difficult at times to follow, and I had to stop many times to look for the path through the high grass. Clearly, the amount of foot traffic here was far less than where I had been for the past two days.
 Overall, the path took me through an interesting variety of landscapes, some that I had not previously encountered. I crossed several country roads, passed by an old cemetery, and near the Baptist Church, I found piles of hospital clothes left in bushes on the side of the trail. My imagination was going wild attempting to come up with a logical explanation for that one. Eventually the trail brought me close to the Elk River, where I followed it for a mile or more. I passed several hikers who seemed to be stopping frequently to watch birds and other wildlife. I continued as the trail lead me up past some beautiful waterfalls, including Mountaineer Falls; a dramatic cascade over steep Precambrian granite ledges.
 I stopped at the shelter by Mountaineer Falls, and it was perhaps my favorite of all that I had seen so far. It had a high steep roof, with a large loft inside with plenty of room on the first floor for perhaps a dozen people. It was clean and well-kept and there were even decorative wreaths hanging in the open-air food prep/dining area. While there, I re-taped several toes that had been bothering me, and very soon, I began to hear the rumble of far-off thunder.
 Later that afternoon, I hiked for about 2½ hours in heavy rains, which greatly increased the flow of all the streams and waterfalls that I came across. When I arrived at Hardcore Cascade, the little mountain stream was now swollen and raging. It was a challenge to get across, hopping from boulder to boulder. Over the course of the day, I must have crossed about 50 mountain streams and springs. I finally arrived at Moreland Gap at 6:15. Because of the continued threat of rain and chillier than usual temps, I decided—for only the third time since beginning the AT—to sleep in the shelter for the night. About an hour after my arrival, Poncho Mike, who had been at the hostel last night with us, rolled in and tented nearby.

Day 32
18.4 miles

Day 33 July 8, 2020 Wednesday
Moreland Gap Shelter, TN to Pond Flats, TN

I woke up late this morning having slept in the shelter last night or—more to the point—not having slept at all. I was wide-awake all night listening to rodents running around getting into everything I had laid out. Never again! This was not the best way to start my day. I noticed that Poncho Mike's tent was quiet, and he apparently was still sleeping. He had told me earlier that he preferred sleeping in late every morning and hitting the trail each day between 10 and 11 AM. Because he was younger and apparently has a much faster pace, we often would finish our days close to the same time in the afternoon.

Despite having been tormented by mice and who knows what else last night, the sun was shining and it was looking like a perfect day. Most of the morning consisted of idle rambling through picturesque forest on nice trails—generally downhill. Off to the east, there were many trails leading to scenic waterfalls, but until I reached Laurel Fork, most of the trail followed a ridgeline. At Laurel Fork, there was a parking lot and many more tourists and dayhikers were now on the trail to another falls, just two miles ahead on the AT.

Perhaps due to my lack of sleep, I stopped more often and longer to talk with people I encountered. There was a long flat stretch of trail, which followed a gravel road with glimpses of Laurel Fork gorge off to the right for several miles. Eventually the path led me deep into the gorge, where a long series of steep Cambrian quartzite steps brought me down to river level and for an eye-level view of the scenic falls. I thought about joining several other folks already swimming over by the falls, but frankly, I was in a hurry to get to my final destination on Pond Flats for the night.

While I was still down by the Laurel Fork River, I ran into Bonanza. He told me his dad was ahead of him and I must have just missed him. I told him about my plan to leave from Damascus in a couple of days, and I sadly would not likely see them again since on my return, they would be about 10 days ahead of me. I remember asking him about his hike down from Pond Flats Mountain, and his comment was "It was brutal." I cringed inside upon hearing that response, but in a way, it motivated me to rise to the challenge.

I continued across the road toward to the start of my ascent and began the long climb to the top of Pond Flats Wilderness. I was still in those half billion-year-old quartzites, and it was steep, very rocky, and narrow in places, just as Bonanza had suggested. However, when I arrived at the top, it didn't take me long to find a spacious flat campsite. There were lots of great options, and I found a perfect site where I watched the sunset from inside of my tent. Before turning in for the night, I rewarded myself with a tasty dinner of Mountain House Fusilli Pasta with Italian Sausage in addition to my standard tuna ramen noodle soup.

Day 33
11.8 miles

Day 34 July 9, 2020 Thursday
Pond Flats, TN to Vandeventer Shelter, TN

I slept well following that awful night stay in the mice-infested Moreland Gap Shelter two nights ago. I intentionally got up late, knowing I was well ahead of schedule, for my planned arrival in Damascus in just a few days. I had hot breakfast of oatmeal and coffee, and because I was feeling wild and crazy, heated up some spam as a bonus. Before breaking camp for the morning, I walked 100 yards up trail to investigate the strange depression next to my campsite. It apparently was the namesake pond at Pond Mountain. The concave forest floor here was totally dry now, but clearly a pond during the spring rainy season. The origin is a mystery to me, no karst here on this mountaintop and certainly no glaciation this far south. Typically, we don't find lakes and ponds on ridge tops in unglaciated terrain. I continued to ponder this (with no conscious intention of making that into a pun), and finally hit trail around 9 AM.

It was an enjoyable and leisurely hike to the bottom of the mountain. There were some blueberries and raspberries beginning to ripen and the bears would soon be on the trails. When I got to the bottom around 11 AM, I headed in the direction of Boots Off Hostel for a cell charge and a short resupply. I ran into some really great people while there. John from Pittsburgh, who was working the store and checking in hikers, was a very friendly fellow and we chatted for a awhile about living in Pittsburgh's North Hills. Poncho Mike, who I had last seen at Moreland Gap and Mountain Harbor, was also staying at the hostel and we caught up for a few minutes. On my way out, I met two young thru-hikers from Mammoth, California: Sand Bag and Dreamcatcher. Later, on my southbound hike while in New England, I would cross paths with them once again.

By the time my device was recharged, it was already after noon, giving me a bit of a late start for the remainder of the day's hiking. I left Boots Off and walked down the driveway toward Lake Watauga. The path for the AT around the lake was delightfully flat and unencumbered by roots and rocks. So much so, that I was even able to walk more than two or three steps without looking down at the ground. This was truly amazing—I was actually able to enjoy the scenery and hike the AT at the same time! At last, here I was multi-tasking, and just like that, my toe caught a small stump of a root, no more than half an inch from the surface of the path, and I found myself preparing for a hard and unexpected landing nearly face first, half catching myself with a flat hard stomp on my already sore left foot. A sudden flash of white-hot pain shot up through my heel, and once on the ground, I remained in place replaying in my head what just happened for a good 20 seconds. So much for not looking at the trail while hiking—another lesson (re-) learned.

I was told that there were some local fishermen who had been, for a while, tossing fish carcasses and guts into the woods behind them when they cleaned their day's catch. Bears, who of course like the taste and smell of fish, had evidently become somewhat used to this wonderful new welfare plan, and appeared regularly expecting to find food from any humans who happened to stroll into their feeding grounds. As a result, there were a multitude of posted signs warning hikers not to as much as stop for several miles along the shoreline due to this dilemma. The old sturdy shelter along this route was even closed permanently, reopened for a short while, and then with the immediate return of the bears, finally dismantled and removed. (In addition, I heard that the poor habits of hikers and campers were also partly responsible for creating this predicament).

Following my lakeside ramble, the trail slowly returned to a steadier uphill challenge for the next several hours. Some sections were very steep, much more so than I had expected, and generally followed the tough Cambrian-age quartzite ridgeline along Iron Mountain. This quartzite bedrock is part of the upper Unicoi Formation, which contains thick beds of coarse grained to conglomeratic sandstones, resulting in these durable ridge forming rocks. As I followed this ridge trail, I caught many glimpses off to the east but the thick summer foliage obscured most of my views.

I finally arrived at Vandeventor about 7 PM; what a disappointment. There was no table, no flat tent site except one under an enormous dead oak (no way!), and only one cramped flat spot directly behind the shelter. The two minimum requirements I have for a suitable tent site are: 1) it must be flat, and 2) there cannot be dead limbs hanging over me. Simply put, I need to be able to sleep, and while doing so I don't want to be clobbered senseless by a "widow maker." I finally decided on the tight flat spot behind the shelter, which was next to the massive northwest dipping Unicoi quartzite—apparently the steep southeast limb of a syncline, and the reason for the long, straight ridgeline along Iron Mountain. Functionally, this campsite fit my utilitarian needs but aesthetically it was dreadful. I made dinner, washed up, hung my food bag, and hit the hay by 9:30. With three days left, I only needed to average 10.8 miles daily to reach Damascus by July 12.

Day 34
12.2 miles

Day 35 July 10, 2020 Friday
Vandeventer Shelter, TN to Double Springs Shelter, TN

I woke to a red sky in the east with a near full moon overhead. *Red sky in morning, sailors take warning.* Earlier in the evening, two bright stars were right outside the window of my tent—likely the planets Jupiter and Saturn. I could see them each time I woke up during the night. I broke camp after a breakfast of coffee and oatmeal, and I was on trail by 7 AM. Sometime around midmorning, I passed a side trail leading north to a large stone monument. I walked over to see what it was all about and soon I was gazing at a grave monument for a mountain man by the name of Uncle Nick Grindstaff, who was buried here in 1923. A sad, telltale inscription on the stone face of the monument read, *"Lived alone, suffered alone and died alone."*

About a mile or so before Highway 91, the trail changed. Instead of following a long straight ridgeline, I began noticing the trail making frequent turns and losing elevation to flatter, less difficult terrain. Here I was crossing a right-lateral shear zone, where northeast trending faults and folds were are abruptly cut by what is known as a cross-strike fault, or discontinuity (CSD). A right lateral shear zone is simply a fault zone where one blocks slides to the right relative to the other block. This type of fault is actually very common in the folded Appalachian Mountains, and in my past work, I had frequently observed, identified and mapped many of these in the subsurface.

I was really enjoying my hike all morning, but I had not seen anyone else all day until I passed Highway 91 that afternoon. On part of the flat leading toward my evening destination, I encountered a day-hiker who told me that he really wanted to do the AT with his girlfriend but she just was not interested. I informed him that most thru-hikers go solo, so why not just do it yourself? Probably not the advice he was looking for. Later around 4:30, I finally made it to Double Springs Shelter after slogging through the rain for much of the afternoon.

When I arrived, I instantly recognized Sam from meeting her on the trail a few days earlier. Sam was a sweet young lady from South Carolina with a sunny disposition just doing select short hikes each day and enjoying the solitude of the southern Blue Ridge Mountains. She camped each night in a hammock, often in remote sites where setting up a tent would not be possible. While we made our dinners together at the camp picnic table, she told me about her work as a produce manager in Greenville at the Swamp Rabbit Grocery Store. I told her I thought Swamp Rabbit would make a perfect trail name for her, and she agreed wholeheartedly.

At 5 o'clock, I was still finishing up my ramen noodles when the skies turned black and the wind picked up. I had pitched my tent down the path near the confluence of two springs, so I quickly grabbed everything and ran down there getting everything secured and inside just as the skies opened up and began a heavy downpour. It was a torrential rain for about 45 minutes, and then it cleared up. When I returned to the shelter a little while later, Swamp Rabbit had left, but Sandbag and Dreamcatcher just arrived—apparently hiking through the driving rain. We talked for a while and then around 8 PM, I headed back down to my tent to attend to my nightly reading and journaling.

Day 35
14.5 miles

Day 36 July 11, 2020 Saturday
Double Springs Shelter, TN to Stealth campsite two miles past Abingdon Gap Shelter, TN

    I woke at 5:32 AM to heavy fog in the little valley where I had set up my tent. I could hear birds and several crows all around me. This would be my last night in Tennessee on the AT. I packed up, brought down my food bag, made oatmeal and coffee, and was on the trail by 8:15. The morning started out cool and foggy and I didn't see anyone for several hours. Later in the morning, I ran into three trail runners and several day-hikers. I never did see where Sandbag and Dreamcatcher had camped last night, nor was there any sign of Swamp Rabbit and her hammock.

    I had my second breakfast at McQueen Knob Shelter, which was quite interesting as it is one of the oldest on the AT. An old log structure built in 1934, McQueen Knob is tiny by today's standards with a dirt floor and barely enough room for two people inside, and very rustic in appearance. I hiked for about an hour more when I arrived at Abington Gap Shelter where there was a large grassy area giving the appearance of a front lawn. The adjacent camp area was very well-kept, and in the warm sunlight, made for a nice place to have a break and just sit and relax for the next 15 minutes. While there, I ran into another hiker who was just coming back from fetching water down the long blue-blaze trail to the spring. He told me that he spotted a mother bear and her cub about 40 yards down the trail. I walked down the trail with anticipation of finally spotting my first bear but saw nothing—par for the course for me! He also told me about some nice campsites about three miles ahead. I decided to hike a little further and camp in one of these stealth sites.

    I was originally planning to camp at Abington Gap because I didn't know if there would be any good campsites within a reasonable distance and I only wanted to do a few more miles for the day. After hearing his report, however, I hit the trail with renewed enthusiasm. On the way, I met a section-hiker from Saylorsburg, Pennsylvania who was down in this area on his way to the Smokies. He completed all of Pennsylvania and much of New England so far. His trail name was Acorn and said he would be hiking through Maine in September. I told him that was the same time that I would be there, and we agreed we would look for each other.

    I found a great campsite about 4:30, right along another ridge top. It was flat and spacious, nestled between mature hardwoods and hemlocks, with a cool breeze blowing across the top. It would've been nice to have others to share this with but for tonight it was just me.

Day 36
10.3 miles

Day 37 July 12, 2020 Sunday "The Road to Damascus"
Stealth campsite two miles past Abingdon Gap Shelter, TN to Damascus, VA

    I woke up a bit later today, about 6:30 AM, with a cool breeze and mountain air helping me sleep well in the early morning. Through the trees I could see the orange glow of the sun just coming up. Today I had only a short hike to get to Damascus, my ultimate destination to complete the first leg of my journey. I packed up, made some hot breakfast and coffee, and hit the trail by 8 AM. From this location, it was mostly downhill to Damascus, but as was so often the case, there were several surprisingly steep uphills too. I was still in the metamorphics of the Blue Ridge, with an abundance of tough quartzite bedrock beneath my feet much of the time. Eventually I made it to the Virginia line, and from there, it was just 3 1/2 miles downhill, which went very quickly.

    On the way down, I collected all sorts of "trail treasures" for my 3-year-old grandson Ezra, such as small rocks, pinecones, moss, and ferns. I arrived at Damascus about 11:40 AM following the white blazes down Main Street to the Broken Fiddle Hostel. This was right along the AT and at the entrance there was a mural painted by Treehouse, the manager/proprietor. I introduced myself to Treehouse, he checked me in, assigned me a sizeable private room, and went over the ground rules.

    There were several others already there including Poncho Mike and a few hikers from Pennsylvania. I met a young man from Philadelphia who was somewhat of a mushroom guru. We talked about the Chicken of the Woods mushroom that I had picked for dinner one night but luckily did not eat. I showed him a picture and he explained that this was not an edible variety, and it's a good thing that we did not eat it. He said it may not have been poisonous but could have made us very sick.

    After getting everything put away in my room, I took a hot shower, did my laundry, and then went across the street to the Marathon convenience-store grill and ordered two freshly grilled greasy cheeseburgers. They just might have been the best two cheeseburgers I have ever eaten. Later I went to Subway for dinner, then returned and hung out in the open-air back porch with the other hikers for the remainder of the evening.

Day 37
7.5 miles

Day 38 July 13, 2020 Monday
ZERO day at Broken Fiddle Hostel, Damascus, VA

    I woke up in a bed this morning, came out to the kitchen for some coffee, and met several other hikers and we all made breakfast together. Today was a ZERO day for me. Because I had arrived an entire day earlier than my original plan, I had all day just to rest before leaving tomorrow morning for the airport and a return to Pittsburgh. This was a great opportunity for me to just rest my legs, feet, and toes for the first time in a couple of weeks. I spent the day leisurely walking a couple of blocks into town to resupply for my return to the trail in 10 days, and to purchase a new pair of Merrell hiking shoes. I also wanted to buy a lightweight sleeping bag to replace my 20° goose down bag which I didn't need in the summer heat. I also needed a new pack, to replace my 6-pound L.L. Bean pack, but could not find the one I was looking for. I would just have to wait until arriving in Pittsburgh to purchase a new one.

Most of the day, I just sat around and rested, and that evening several other hikers arrived including Beauty and the Beast, Sorrel, Bells, Swamp Rabbit, and Honey Mustard. That evening, Honey Mustard, the chef from Boulder Colorado, made dinner for everyone on the outside charcoal grill. He treated us to grilled chicken thighs as well as pork, chicken, and beef shish kebabs. I can't even describe how incredibly delicious all of this was and we feasted well into the evening. In addition, Swamp Rabbit's boyfriend, an accomplished musician in an internationally traveled band, provided on-line music for us from their most recent concert performance in Belgium. This was a great way to end the first leg of my journey. The next morning, I arose at 4 AM and was picked up by my shuttle driver Al, who drove me to the airport for my flight to Pittsburgh.

Day 38
0 miles

Getting started! Posing by the famous Archway marking the start of the approach trail at Amicalola Falls

Coming down from Tray Mountain early morning after several days of rain
Flame Azaleas and Mountain Laurel are in bloom everywhere

Little Bald Knob campsite on ridge top. This was the first of many summer ridge-top camps affording a cool breeze, filtered sunlight and great views

Highly contorted quartz-filled veins in Precambrian metasediment

Views from trail from Albert Mountain on way to Long Branch Shelter

Beautiful sunset looking west from Siler Bald following afternoon hailstorm

My first break after 11 days on the trail. Relaxing on the river at the NOC.

Running into my bi-weekly buddies Rocky and Bonanza near Newfound Gap

Spectacular scenery looking west into Tennessee from Charlie's Bunion

Finally resting my sore, tired feet on Max Patch Mountain, North Carolina

Beautiful sunset view on Max Patch Mountain, North Carolina

Six-day resupply while at Laughing Heart Hostel, Hot Springs, North Carolina

Set up for the night near Street Gap overlooking Big Bald Mountain at sunset

Hitting the trail at sunrise on way to Big Bald Mountain

Sunrise from inside my tent at Beauty Spot, North Carolina

A morning hike along the ridgeline balds toward Hump Mountain

Arriving in Tennessee! Trail coming down from Hump Mountain through Doll Flats

An ominous sky greets me after leaving Mountain Harbour Hostel

Chef Honey Mustard preparing a feast for all of us on our last night at the Broken Fiddle in Damascus, Virginia

One of the many rhododendron tunnels encountered along the trail in NC

Part Two: Endless Blue Ridge Summer

Chapter Six
Return to the Trail

Day 39 July 23, 2020 Thursday
Damascus, VA to Stealth Campsite near Cuckoo Knob, VA

    I left PIT 10:30 AM, arrived Tri-Cities at 2:30. After retrieving my pack from baggage claim, I walked outside and looked for my shuttle. My driver was Frieda, Al's wife, along with their granddaughter Kaylee, who quietly sat in the back seat for the duration of our trip. Frieda was very friendly and talkative, and we both engaged in lively discourse the whole way to Damascus. When we arrived at town, we first stopped at Sundog Outfitters so that I could buy an 8-ounce gas canister and a few other incidentals. Next, she dropped me off at the steps to the AT trailhead on the Virginia creeper trail along US Route 58.

    By now, it was about 4 PM and I was organizing my pack along the road prior to entering the trail. I noticed the sky becoming increasingly darker, as if someone was slowly turning down a dimmer switch on the lights. I instantly knew what was coming and hastily put on my raincoat before throwing on my pack. I started up the trail in this strange twilight, and within 10 minutes the heavens opened, and it started to rain as hard as I think I have ever seen. The rocky trail beneath my feet soon became a turbulent cascade of muddy water, pooling up into little lakes in all the long flats between the steeps. As I continued to climb higher, all of the steeper sections were quickly being transformed into raging waterfalls. There was nowhere to step in order to avoid soaking each footstep in a barrage of muddy water. This heavy downpour continued for the next three miles, where I finally found a suitable campsite near the top of Cuckoo Mountain.

    Through the curtain of falling raindrops, I could see this was a nice clearing with a fire pit surrounded by several large logs and ample space to put up my tent. There was even a backdrop of giant granite boulders, all covered in a shimmering green collection of moss and ferns. I hurriedly put my tent up in the rain, careful to keep the contents of the rest of my pack as dry as possible under these monsoon-like conditions. I eventually fired up my Pocket Rocket cook stove, and carefully prepared ramen noodles, all under cover of the vestibule of my tent. My raincoat was drenched, as was my shirt, shorts, shoes, and socks. I carefully dried off and entered my tent where I stayed warm and dry all night. Rain continued pretty much throughout the night and into morning. I remember thinking to myself, "Welcome back to the AT!"

Day 39
3.0 miles

Day 40 July 24, 2020 Friday
Stealth Campsite near Cuckoo Knob, VA to Lost Mt. Shelter, VA

The next morning, I had to wait out rain, which was very heavy until about 7:30. By 8:00, it had stopped, allowing me to get out of my tent and put on wet shorts, shoes, socks, and shirt, then make coffee and hot oatmeal to warm up. Temperature this morning was 62°—not too cold but everything was soaked around me. All morning I hiked in misty overcast conditions, but no rain and with nice views from time to time. The AT crossed over US Route 58 several more times this morning, and on one of my crossings, I passed a southbound (SOBO) hiker wearing just one shoe. He had been on the trail for a few days with his dog and explained that one of his heels developed such severe blisters, that it was actually preferable just to hike barefooted on that foot.

Around noon, I stopped for lunch on top of Straight Mountain at Saunders Shelter. I took off my shoes and socks and hung them on my poles to dry. This always seemed to help my feet for another round in the afternoon. After lunch and a short power nap, I met four octogenarians out for a day hike. One lady said she hiked the AT when she was between 57 and 70 in sections with her partner. They all looked to be in great shape for their age!

The Appalachian and Virginia Creeper Trails cross several times just past Damascus, and even follow the same path for short distances. On this section, I encountered mostly bikers, some hikers, and even a few people on horseback. I talked to several other folks on foot, but all the non-hikers were either zooming past me too fast or preoccupied in some other way with their mode of travel. Because the Virginia Creeper Trail was converted from an old railroad right of way, it contained several nicely refurbished trestles crossing over various streams. On this stretch, I crossed over one of the forks of the Holston River, which appeared to be a nice trout stream. There I noticed several fly-fishermen wading in the lively water casting their lines again and again.

While strolling along, I thought about something my wife asked me while I was home the previous week: "What in the world do you think about all day long?" To be honest, I had to stop and consider this before giving an answer because it wasn't entirely obvious. After giving some thought, I replied, "Well, a lot of things, but mainly survival." In other words, a great deal of thought and energy goes into essentially planning and organizing all the things I need to do with my limited resources and time so that I can stay dry, pace myself, reach target mileposts on time, and always have enough food and water. I know it sounds trivial, but it is probably the one thing that never escapes my attention.

My destination for today was Lost Mountain Shelter, and it was only shown to be 6.4 miles after my lunch stop. For some reason, it seemed to take an inordinately long time to do this mileage. The trail passed over and around massive outcroppings of weathered gneiss, somewhat of a change from the quartzites of the past week before Damascus. Finally, around 6 PM, I arrived and there were two men from Georgia standing around the picnic table fixing their dinner. We exchanged greetings and began discussing our backgrounds and our reasons for being out on the trail. I learned they were also retired, having worked for the Department of Defense (DoD). One of them was also into genealogy like me, and we talked at length about that for a while.

I next went about my business of getting water, unpacking, and setting up my tent. While having dinner at 7:30 it began to rain again. Rain continued steadily but I managed to make it to the bear box with my food and back to my tent for the evening without getting too soaked.

Day 40
12.6 miles

Chapter Seven
Mt. Rodgers, Grayson Highlands and Exploding Fish Oil

Day 41 July 25, 2020 Saturday
Lost Mt. Shelter, VA to Rhododendron Gap, VA

I awoke at first light to the sound of birds singing. I then realized it was after 6 AM due to shortening of days. I looked up to see clear skies. It was 62° and shaping up to be a gorgeous day ahead. I came out of my tent and noticed another tent not more than 10 yards away that I had not noticed yesterday. After breaking down camp, I walked over to the shelter to cook breakfast and to talk to the two DoD guys out for the weekend. Soon, Floss emerged from his tent, walked over to say good morning, and explained that he arrived around 10 PM last night. He logged 40 to 50 miles per day and was trying to break the record for AT fastest thru-hike. I was amazed to hear this. This would be like running two marathons every day for about 45 straight days! (Floss was a 2018 AT NOBO finisher, but I have been unsuccessful in finding even a line-item mention of him for 2020).

The hike to the top of White Top Mountain was very pleasant and easy in the cool morning fog, eventually bringing me to Buzzard Rocks. The top of the mountain was a grassy bald with spectacular views in nearly every direction. There I met a day-hiker with two tattoos on his leg: one was a coal miner, and the other was the WVU Mountaineer. We talked for 10 or 15 minutes about our favorite places in West Virginia and about points of interest along the AT. Later I continued north across the bald, following the trail back into the woods where it eventually led me to Elk Garden on Virginia 600. At Elk Garden, I talked to several people beginning their ascent to the Mt. Rogers/Grayson Highlands area.

I made pretty good time on the steep, rocky trail leading to the top, passing about a dozen other hikers along the way, but decided against taking the blue-blaze trail to the Mt. Rogers summit (all treed in, no views). At 5,729 feet elevation, Mt. Rogers is the highest point along the AT in Virginia as well as the highest elevation in the entire state. Geologically, this is a fascinating part of the Blue Ridge province, uniquely containing rock evidence for both volcanism and glaciation from many hundreds of millions of years ago. The volcanic rocks here are remnants of past eruptions, which occurred following the breakup of supercontinent Rodinia as the ancient continents were being ripped apart by plate tectonic movement about 760 million years ago. In addition to the volcanics, evidence for a Neoproterozoic glacial event is also preserved in the rocks here. Lithified glacial till deposits, known as diamictites and tillites reveal an ancient past of Earth's history, known commonly as Snowball Earth.

This was a time in geologic history when perhaps the entire planet was covered in ice, earning this period the chilling name of the Cryogenian. Following this ice-age, there were several more throughout the next several hundred million years—including the last, beginning just 2.6 million years ago, within which we currently reside. Later, my SOBO journey from Maine back to central Pennsylvania would be almost entirely on rocks affected by this latest glacial event.

As I rambled along this highland trail, I spotted a lot of hikers already camped out for the day with their colorful tents settled in amongst the coniferous background forest. I continued past them and on to the Thomas Knob arriving there about around 3 PM. While at the shelter, I talked with two past thru-hikers, Yak and his girlfriend, Snug. Both were from Pennsylvania, and Snug had been a student at Pitt. We chatted about the AT, COVID, considerations for postponement until later in the year, and about some of their favorite parts of Maine and Pennsylvania. Following our discussion, I next walked down blue-blaze trail to the spring to refill my water bottles. There was a white PVC water pipe coming out of the ground, flowing a 2" stream of water into a small clear pool, all of this inside of a fenced in area. The fence was there so that the wild ponies would not get in and contaminate the spring. After getting water, I got back on the trail and was considering going five more miles down the mountain to Wise Shelter. However, it was getting into late afternoon, and after the first half mile, I spotted a nice secluded campsite nestled in between large spruce trees near Rhododendron Gap. From here, I had spectacular views of the mountains off to the east toward Grayson Highlands. One of the hikers I met on the way up told me that he had set up camp here earlier and moved because the longhorn cows were getting a bit too friendly for his liking. I guess I just wasn't worried about cows or ponies after having survived the Smokies and Nantahalas. I found a good spot and pitched my tent.

It was looking as if thunderstorms were on the way, so I had to hurry and get my tent set up and make dinner as quickly as possible. While my ramen and smoked oysters were cooking, I ran over to the bear box to stow my food bag and swiftly returned to get under my vestibule just in time. While the rain poured down on the mountain, I enjoyed my dinner reclining on my sleeping bag and pad, eating under the protection of my trusty Dyneema tent. There were no other options and I had to be extra careful so there would be no spills! It rained very hard for about 90 minutes with lots of nearby lightning and ear-splitting thunder throughout the storm. Slowly, the skies began to clear up around 6 PM revealing blue patches between the departing clouds. Looking around me, I could see there were many clusters of tents in every direction. Because of the rain, I noticed that a small amount of water had come into the tent because my Thermarest air mattress had been up against the rear wall. I was able to easily mop it up with a pair of socks. Earlier in the day, on one of my breaks, I added three strips of Leuko tape to a pinky toe Band-Aid. Bad idea! It ended up taking me 30 minutes to remove the stubborn adhesive from the already blistered skin. Note to self: never apply Leuko tape to pinky toes!

After dinner, I walked around the campground area while watching a beautiful sunset paint the sky from the western setting sun. I spent most of the next hour chatting with other campers including an adventurous mom and her kids from Raleigh North Carolina and two men who had somehow managed to stay dry under a small tarp throughout the rainstorm.

Day 41
13.0 miles

Day 42 July 26, 2020 Sunday
Rhododendron Gap, VA to Old Orchard Shelter, VA

I awoke to first light, heard birds chirping, and the gentle rattling of a soft wind outside of my tent. I was relieved to see a beautiful sunrise with clear skies and no cloud cover. I broke camp and was on trail by 7 AM. As I climbed the rocky summit that I had been looking at the evening before, I noticed there were several longhorn cows and ponies wandering around. I later ran into the four guys who had camped just before the Mt. Rogers approach trail. They seemed in a big hurry and I let them pass me. Several of them were carrying what appeared to be soaked, rolled up tents or sleeping bags, making me wonder about how they survived the cool night following last evening's dinner-hour monsoon. On the way down the mountain, I saw lots more ponies, longhorn cows, and about 10 to 15 people.

I decided to make today a light day in order to recover from feet and other pain. Good call. I got to Wise Shelter by 9:30 AM for a break and was back on the trail by 10:00. I later continued north through balds, meadows, patches of woods, and some highland meadows. The scenery was extremely variable. After I had come down from the highlands, I ran into a group of four young ladies on their way up to Grayson Highlands. We all stopped and talked for a bit, and the discussion led to the ponies and longhorn cows up in the highlands. I mentioned that I couldn't get that song, "Wild Horses" out of my head. One of the girls replied, "thanks a lot, now that's all I'm going to be thinking about too!" And as a pitch back she added, "And I can't get 'It's Raining Men' out of my head!" "Noooo!" I thought to myself!

Around noon, I came down to a road and parking area where there was a fenced-in corral known as The Scales. There were four families packing up to go home, and as I walked past, they called out to me offering me beer, food, and water. Trail Magic! I stopped and talked with them, and gratefully accepted their offer for all kinds of food, snacks, and goodies. They were super nice folks, and I talked with them for about 30 minutes. Several of the kindhearted ladies even packed me a second lunch after I had eaten a sandwich that they prepared for me earlier. I was truly thankful for their benevolence and said goodbye as I left with a large goodie bag for later.

The next several miles were through a thick spruce and birch forest—very rocky and slow, but one of the most beautiful forests I can remember since Unaka Mountain. It, too, reminded me of the spruce forests of Maine. I arrived at Old Orchard Shelter about 3 PM. There was no one there and I set up my tent, dried out all my gear, had my second lunch, got water, and rested for a while. Later, I cooked ramen and chicken for dinner, and afterwards went over to meet three Ohio guys camped out next to me. We talked for a while about my present thru-hike, some of their past backpacking trips, and of course about some NFL-team rivalries. After that night, the three of them were planning to hike back to their car parked about five miles down the mountain on SR 603. They said they had some food left over and asked if I needed anything? I replied that I did need one more dinner, and they graciously presented to me one scrumptious Mountain home stroganoff dinner and several Clif bars. Wow, trail magic twice in one day! It was a short day, but certainly quite enjoyable.

Day 42
10.1 miles

Day 43 July 27, 2020 Monday
Old Orchard Shelter, VA to Trimpi Shelter, VA

I set my alarm for 5:15 AM to get an early start for the day. Sunrise at 6:35 made breakdown of camp all in darkness this morning. I had oatmeal and coffee and was on trail with remainder of coffee in hand at 6:42 AM. I made it down to the bottom of the mountain before 9 AM, and then started up the other side toward Hurricane Mountain. Unbeknownst to me, I had just crossed, for the first time since beginning my journey seven weeks ago, into the rocks of the Valley and Ridge sedimentary section. I was no longer in the relentlessly complex world of those Proterozoic metamorphics of the Blue Ridge province. Somewhere in this valley I had just crossed over a major thrust fault, and I was now trampling on much younger Paleozoic rocks.

About halfway to the summit, I met an older gentleman wearing a straw hat by the name of Wayne, who said he was hiking with his son-in-law. Although Wayne was alone, he said they would be meeting up ahead somewhere later today. We were both going the same direction, but my pace was a bit faster and I soon was far ahead. Later, I would stop for a few minutes, running into Wayne several more times, talking with him for an hour or so and then moving on as our paces dictated. He didn't have a pack, so I guess he wasn't going far. He did tell me that he was a semi-retired bridge engineer from Cincinnati, and that he had 10 stents in his heart. There were some sections along the trail this morning that were very poorly marked, and on several occasions, I backtracked to be sure I had not missed a turn. On the way down the mountain, I spotted an extremely unusual mushroom, which I assumed was poisonous. It was very dark, almost black, and had the shape of a chanterelle with its trumpet-like inverted cap. Upon careful examination, my first naïve and somewhat paranoid thought was "black death." Later research revealed this to be a Black Trumpet mushroom, or Craterellus cornucopioides, which not only is edible, but reportedly one of the best tasting fungi on the trail.

Later in the day, I met several other hikers along the way, including one young man telling me that he didn't have a trail name, but liked the idea of Forest Tortoise. I said to him, "Great! Your official trail name is now Forest Tortoise." He gave me a grateful smile and seemed very pleased as we parted, continuing on our separate ways. Later, on my way to High Point Mountain there was a major detour on the trail. Apparently, the bridge was out at Comer's Falls, so the AT was rerouted up a steep asphalt road for 1.3 miles. This was a long arduous climb on such a hot and humid day, and I wondered if Wayne, who had fallen behind, ever made it. There were several cars that passed me on the long uphill road climb, and I kept looking to see if any of them had a passenger with a straw hat. I worried about those 10 stents in his heart, but I never did see him again.

When I got to the top of the hill, the detour ended and I resumed on the trail, which led me back into the woods near High Point Summit. I arrived at a trail intersection around noon, took off my pack, and had a one-hour lunch with my shoes and socks off. My feet were really hurting today and I took a long rest after lunch. It was only another 3.5 miles to Trimpi Shelter, but the trail conditions were such that it made the afternoon hike very difficult. I finally arrived by 3 PM and decided to call it a day. I set up camp, got water, had an early dinner, and was in my tent by 6 PM. It rained for about one hour while I was making dinner under the cover of the shelter, but then was clear for the rest of the evening. I was totally exhausted after today and I prayed that tomorrow I would feel better.

Day 43
14.1 miles

Day 44 July 28, 2020 Tuesday
Trimpi Shelter, VA to Chatfield Shelter, VA

At 5:45 AM, I woke up to my alarm and could see that it was still dark outside. I packed up my gear, broke down camp, had breakfast of oatmeal, raisins, and coffee and was on the trail by 7 AM. As the emergent sun slowly burned off the morning fog, I passed through several old fields where the openness revealed misty but magnificent views of far-off mountains, many of which were beneath my intended path for the day. This was a well-planned day beginning with a 15-minute stop at 9 AM, and then another around 11:50 AM upon arriving at Partnership Shelter for a long lunch after 10-mile morning trudge. All the while the skies were overcast, threatening rain for several hours leading up to noon. A few minutes after I arrived, the skies suddenly opened up and the rain came down in buckets for the next hour. I was under cover for the full hour, having lunch and resting with my shoes and socks off. At 1 PM, just as the rain came to an end, I threw back on my pack and resumed my hike. The timing here was simply amazing and peering into my future it was looking quite auspicious!

Partnership Shelter was truly unique. In addition to sleeping 16, it also had a shower and a spigot in the back, neither of which was turned on, however, due to COVID. There was a ladder on one side of the wall leading up to a loft. I climbed up to have a look and noticed there were several items lying around as if someone had been staying there. There was also a National Forest Service visitor center right down the trail about 100 yards, and although there were a lot of cars in the parking lot and people working there, the visitor center was locked. I looked around to see if there was anyone outside with whom I could talk but saw no one.

I continued my hike for the next several hours arriving at Chatfield Shelter about 4:30. I set up camp, looked for a good place to hang my food bag, had dinner, and was in my tent by 7:30. For the first time I can remember, I saw no one on the trail the entire day. All afternoon and into the evening the clouds were breaking up and the sun was peeking out from time to time. The day ended with clear skies and a beautiful sunset.

Day 44
16.8 miles

Day 45 July 29, 2020 Wednesday
Chatfield Shelter, VA to VA 42 (Bear Garden Hostel), VA

My alarm went off at 5:45, but I kept hitting the snooze until about 6 AM. I began my morning ritual of packing up and organizing everything inside my tent. When it was time take my morning vitamins and fish oil capsule, I noticed there was no more water left in my water bottle. Rather than choking down the dry pills, I put the vitamins and fish oil in my left pants pocket. When I eventually walked over to the picnic table, I reached in to retrieve the vitamins, and found the fish oil had exploded, saturating the whole front of my shorts and left leg. Just what I needed, smelling like a fish and attracting more flies, gnats, and bears! Certainly not the best way to start my day.

Around mid morning I crossed over the great Pulaski fault system, a northeast-trending thrust fault that extends from North Carolina into central Virginia roughly along the eastern edge of the Shenandoah Valley for several hundred miles. I was now in the Cambro-Ordovician carbonate rocks of the Shenandoah Valley near the I-81 underpass and was enjoying a slice of pizza from a local convenience store. Soon another hiker strolled over from the southwest and we both chatted for a while. Spokes was out for several days and planning to do about 20 miles each day—a rather aggressive schedule, I remember thinking at the time. He said he was hiking to Knot Maul Shelter for the night, and I responded that I doubted I would make it that far. He was in a hurry, so we said our goodbyes and parted. I returned inside to buy supplies for the next few days as well as two more slices of pizza and a large Mountain Dew.

Following my break, I continued following the white blazes under I-81 and across several fields and small hills. Eventually I reached an opening in the woods where the trail became less gentle and finally started gaining elevation upon reaching the Devonian Chemung. These rocks were nowhere as tough as the Silurian orthoquartzites I would experience on nearby Walker Mountain later that day. By early afternoon, I was on my way up and over the smaller Devonian peaks that eventually led up to Little Brushy Mountain. It was a gorgeous summer day, and the sun was heating up the air and rocks along the trail. About 2/3 of the way down the mountain, I encountered a very calm, but quite hefty, five-foot timber rattlesnake right along the left side of the trail. I stopped dead in my tracks and waited for a few seconds while the snake remained totally still. This was certainly not something I would want to step on, and I waited a few seconds more! The good thing about these pit vipers is that they are so big, you really can't miss them. All day long, I would be looking down at the trail directly in front of me, and something this size would be almost impossible to miss. I gave him a wide berth; walking well off the trail and around him; watching him all the while as he slowly slithered away into the high weeds. I wondered if he could smell that nasty fish oil on my left pant leg?

When I made it to the bottom of the mountain, I stopped at the Crawfish Valley Campground, and for the second time today, attempted to wash out the fish oil smell from my hiking shorts. I even tried using some of my biodegradable liquid soap—something I rarely opened in my five months on the trail. It seemed like the more I scrubbed, the oilier my hands and everything that I touched became—including my trekking pole handles and my water bottles. Later, I would learn that this campground was the site of that dreadful machete murder last year on the AT. I continued up and over the next hill, running into several other hikers along the way. I had to face it. I smelled like a fish and was getting riper by the minute.

By about 6 PM, I finally emerged from the woods into a field where the trail intersected a country road near the little town of Ceres, Virginia. As I approached the road, I looked to my right, and there was a sign announcing, "Bear Haven Hostel, 0.2 miles ahead." I was aware this hostel was somewhere nearby, but I had no idea it was right in front of me. This was just what I needed to clean up so that I wouldn't have to crawl into my tent tonight smelling like a fish for all the bears. When I arrived at the main bunkhouse, I knocked on the door and there was only one other hiker inside, a young lady from Louisiana who was sectioning hiking for a couple of weeks. Brandy explained that Bob, the owner, would be coming back around soon, so I waited for a few minutes until he arrived. That evening I was able to shower and do laundry, thereby eliminating the fish-oil problem that I wrestled with all day long. We talked for a while, and later I feasted on a couple of microwave sandwiches for dinner along with several Yoo-Hoo chocolate drinks from the refrigerator in the bunkhouse.

Day 45
17.2 miles

Chapter Eight
Virginia's Valleys and Ridgelines

Day 46 July 30, 2020 Thursday
VA 42 (Bear Garden Hostel), VA to Davis Farm Campsite on Garden Mtn., VA

    I had a poor night sleep in Bear Garden Hiker Hostel, probably because of that can of soda and two bottles of Yoo-Hoo so close to my bedtime. Waking at 6 AM, I had coffee and toast with some of the delicious homemade jams made by Bob's wife and was out the door and on my way to the trail by 7 AM. The mild weather and trail conditions enabled me to make good time in the morning, rambling up and over Brushy Mountain and some smaller hills. By 11 AM, I was on my way up to Chestnut Knob.

    I had been looking forward to getting to the top of this mountain for a long time, so that I could look down into Burke's Garden from the top. There was only about one hour's worth of a real grind up the steepest part of climb, and from there it was gradual and not at all difficult—just long. I arrived at a spring pond about 12:31 PM, filtered water and filled my bottles, and then made it up to the shelter by 2 PM. As I climbed higher, there were frequent spectacular views off my right shoulder to the southeast. The last several miles were mostly grassy balds, and soon I was at the top of Chestnut Knob overlooking Burke's Garden.

    Spokes was just getting ready to leave when I arrived. I tried explaining to him about the incredibly interesting geology and how Burke's Garden is such a perfect example of a doubly plunging breached anticline. A breached anticline is simply a large, upside-down U-shaped fold, with its crest (or center portion) eroded to expose older, generally less-resistant rock layers. He looked over at me and said, "You aren't a geologist, are you? That makes two of us!" It turns out that Spokes, or Harry, went to University of South Carolina for his BS and University of Kentucky for his MS, both in geology. He then went on to work as an environmental geologist for most of his career. We jabbered on for a few more minutes about the rocks and trails. Soon after that, he departed announcing that he was going to try to make it all the way to the shelter at Jenkins by evening.

    For the remainder of the afternoon, I tried my best to catch up with Spokes or make it to Jenkins, but it was too late in the day and I just couldn't do it. The skies were becoming increasingly overcast, with occasional heavy rain showers throughout the afternoon. The geology along Garden Ridge, however, was fascinating. All along the ridgeline was the gently dipping southeast limb of Chestnut Ridge Anticline. The Silurian-age Clinch Sandstone here is several hundred feet thick and the escarpment formed on the northwest edge is dramatic and very abrupt all along the ridgeline. The trail was continuously tedious because it repeatedly led me up and over massive orthoquartzite boulders where I would walk for a while and then need to climb back down. I was hurriedly following the path wherever it led, but generally at a very slow pace because of the variable terrain. Despite the difficulty, it was fascinating, and I really enjoyed the challenge. It just took a really long time. I noticed some Silurian trace-fossil beds for which I couldn't recall the name. Trace fossils are imprints such as burrows, trails, or footprints left behind by an animal and preserved in the rock record. I contacted Lee Avary, a retired geologist and good friend from the West Virginia Geological Survey who reminded me they were Arthrophycus trace fossils, either resulting from the tracks of arthropods (trilobites) or annelids (worms) on the sandy bottom of a near-shore Silurian seabed. The debate about the kind of organism making these traces in the sediment continues today.

I resumed hiking through the intermittent rain well into the late afternoon. There were no places to camp because of how uneven and boulder-strewn the ridgetop was. I plodded on and eventually arrived at the Davis Farm Campsite. There I quickly set up my tent just before another round of intense rain. The relentless deluge forced me to cook my ramen noodles and eat my dinner under the protection of my vestibule, while the rest of me stretched out inside of my tent.

Day 46
18.7 miles

Day 47 July 31, 2020 Friday
Davis Farm Campsite on Garden Mtn., VA to
Helveys Mill Shelter, VA

I awoke to continued wind, fog, and overcast skies. It looked like rain, but never did all day. Breakfast bars and cold coffee, and I was on trail by 7:45 AM with a long downhill to shelter, privy, and water, then I was on my way. I had a good pace all morning coming down Brushy Mountain to Laurel Creek Falls where there was a nice new bridge across the creek. Lots of kids were swimming and playing with their moms in the clear open pools below two levels of waterfalls. I sat on a rock near the lower falls and soaked my feet in the cold running water for 10 to 15 minutes while I enjoyed my lunch. It felt great! While I was cooling off my toes, one of the slightly plump moms slipped on the rocks and plunged into the water, prompting stares and giggles from the children. She seemed really embarrassed, but clearly wasn't hurt—at least not physically. I later wallowed back over to the bridge and found a comfortable place to sit by the stream next to some rocks and took a quick power nap. After about five minutes I was totally recharged.

That afternoon was long but I was able to keep up a better than average pace of two-and-a-half to three miles per hour on a good path through mountains, compared to my average pace of just two miles per hour. Large sections of the trail were on gravel roads and not all were well marked, causing me to backtrack several times. At the bottom of the hill, I looked ahead and noticed a large sign advertising Brushy Mountain Outpost and Deli. The sign said, "open always" and "welcome hikers." I made a beeline for the little house where the sign directed me only to find a locked door—probably once again due to COVID. I crossed over I-77 and then passed by some spectacular Devonian outcrops along the road showing nice asymmetric folding patterns in the thin sandstone units.

I recall in the last year when I would take my daily hikes around North Park near my home in Pittsburgh, I would consistently average three miles per hour. I generally hiked with one half-liter bottle of water, and no daypack. When I began back in June on the AT with a 40-50 lb. pack, my average dropped to two miles per hour. So today was a nice pace and a change from the norm.

Eventually I made it all the way to Helveys Mill by 5:30. My water source was way down the hill, probably one of the longest water runs I had to date. As always, I was absolutely ravenous by the time I set up camp. The dinner I made on this night was awful, consisting of cut up pieces of semi-old beef sticks, fried, and then boiled with ramen. It was gross. I know I shouldn't have eaten it, but because I was so hungry, I did, and hoped it would not make me sick. For greens, I also mixed up a cup of cold green slime from my Ghost Green powder mix. As bad as that might sound, it was actually delicious. Afterwards, I retired to my tent and spent about two hours reading, reviewing, planning, documenting, and wondering if I was going to barf. Finally, about 10 PM, I tried to get some sleep.

Day 47
17 miles

Day 48 August 1, 2020 Saturday
Helveys Mill Shelter, VA to Kimberling Creek
Suspension Bridge, VA

I awoke to first light, had cold coffee, and a couple energy bars for breakfast, and was on trail by 7:00 AM. Just before hitting the trail, however, I moseyed over to the privy only to find a large COVID warning sign and padlock on the front door. Most of the shelters and privies on the AT had warning signs either on one, the other, or both. Like most hikers, I always carried hand sanitizer to use before and after privy and shelter visits. It always seemed to me a bit of an overreach by the Appalachian Trail Conservancy (ATC), local trail clubs, or the NFS to padlock the doors to a facility that is miles from civilization and only lightly used at best. Doing so was not going to deter hikers from using the trail. However, it might increase the amount of unwanted human waste left by those not using cat holes responsibly.

For some reason, I had trouble establishing any real momentum on this day. Terrain was not overly difficult, but the trails all morning were following ridgelines over Devonian Chemung mountains. Within the Chemung, there are primarily two types of rocks: sandstone and shale. The sandstones range in thickness from a foot to tens of feet, generally bounded by less resistant shales of similar thickness. This geologic predicament makes foot travel a bit different than the long continuous flats of the Silurian capped ridgelines trod upon this past week. Although both are sedimentary, the Silurian age sandstones are generally much thicker and are often orthoquartzites, meaning pure quartz sandstones with the individual sand grains cemented to one another by quartz cement. These are very resistant to weathering and represent some of the "toughest" rocks along the AT.

In contrast, here in the Chemung I was continuously rising and falling in elevation due to the less resistant rock type more prone to weathering, where these sandstones are comprised of quartz and feldspar grains, and the grains are cemented by calcium carbonate, a much weaker cement. In my past career, I drilled countless wells into these same sandstones in Kentucky, Virginia, West Virginia and Pennsylvania in search of oil and natural gas. As I plodded along the well-blazed trail, there were lots of ups and downs along these ridges, making it an exhausting hiking experience.

I made it to the next shelter for lunch by 12:30. It was delightful weather all day, 65° in early morning up to 75° by noon with plenty of sunshine filtering through the trees overhead. The day warmed up through the afternoon with very slight drizzle but no significant rainfall. I only ran into one other person in the morning, and literally not a single person in the afternoon until I got to Kimberling Creek footbridge about 5 PM. There I met Fungi, evidently named for his vast knowledge of mushrooms, which he routinely foraged during his daily hikes. He graciously invited me to join him for the evening at his little campsite by the river, and I gladly accepted. Fungi had a nice campfire going next to his tarp-covered hammock, and I found a suitable spot nearby to set up my tent.

Over the next several hours, we had dinner together, talked a lot about our past adventures, and I learned a lot about mushrooms as well as his vast knowledge and experience with handling rattlesnakes he finds along the trail. He showed me his latest collection of Chanterelle mushrooms that he had harvested along the trail, but explained that for many edible fungi, there are often poisonous look-alike varieties. He went on to say that eating the wrong mushroom could result in a trip to the hospital for an organ transplant. Fungi didn't offer me any of the Chanterelles and I didn't ask for a sample. He also revealed to me how, many years ago, when he was working on a construction project with his brother, he had an unfortunate nail gun accident that caused him to lose one of his eyes. Fungi was a very friendly and interesting person, and I was glad to have had the opportunity to spend time getting to know him.

Later that evening, two fishermen we had seen earlier on the river returned to cross over the footbridge. I said hello and asked them where they were from. They responded, "Wise County, Virginia." I told them that I was very familiar with Wise County having worked the Roaring Fork and Nora oil and gas fields for many years as a geologist for EQT. It turned out they knew some of the employees from our Big Stone Gap field office, including Junior Shupe. Small world! Fungi and I talked for a while longer until about 9 PM, when I finally retired to my tent for the evening.

Day 48
17.1 miles

Day 49 August 2, 2020 Sunday
Kimberling Creek Suspension Bridge, VA to Doc's Knob Shelter, VA

At 5 AM I woke up in the dark, packed up my gear and was on trail by 6:30 AM. For breakfast, I had cold coffee and ate two energy bars while beginning my hike for the day. The morning was very flat and easy hiking, and I was able to make good time, moving along between two to three miles per hour. I met several hikers along the way, including two girls with a small dog. Last night, Fungi highly recommended I stop at Dismal Falls, but when I arrived at the blue-blaze trailhead, I decided that a 0.6 mile round trip was too much extra time. It was beautiful weather all day; sunny with a cool breeze and actually very windy on top of several mountains. There were many lovely views along the way. I noticed that the Clinch Sandstone was dipping back toward the southeast on many of these northeast trending ridgelines. I mentioned later to someone that when hiking along these ridges you can always tell on which side of the ridge is the anticline and on which side is the syncline, to which I received a blank stare.

After several more hours of ridge hiking and passing by several more hikers, I arrived at Docs Knob by about 5 PM. This uniquely designed structure was down by a spring and nestled in a grove of mature rhododendrons. It had an attached front deck upon which were several picnic tables and benches. Ironically, I met two North Allegheny High School (NA) grads from 2011, the same high school from which my son and daughter graduated a few years earlier. Both hikers were veterans; one had served in the US Air Force, and the other in the Marines. They had tons of food and spices, all spread out over the table on our spacious front deck. More ironically, they even had been sautéing Chicken of the Woods mushrooms, which I finally had the opportunity to try for the first time. Olive oil, garlic, and other spices were added for exquisite flavor and I thought they tasted like lobster. The ex-Marine also used to be the head chef at a popular North Hills restaurant, "Blue." His trail name was Food Truck because of all the food he carried in his oversized pack. We all talked a lot about Pittsburgh, North Allegheny High School, and local restaurants.

After a while, a fourth hiker arrived, Shutter from Florida, named because he wore a 35mm digital camera around his neck to photograph nature and scenery. Shutter also wore a tracking device that allowed his family and friends back home to always know his whereabouts. After about an hour, he informed us that a family member located him via this device and was going to be arriving in a few minutes. Sure enough, his uncle soon arrived and away Shutter went for an unexpected night out to a real restaurant! The remaining three of us enjoyed ramen noodles, mushrooms, and an assortment of other delectable, dehydrated food products for our dinner.

Day 49
17.6 miles

Day 50 August 3, 2020 Monday
Doc's Knob Shelter, VA to Lane St, Pearisburg, VA
(Angel's Rest Hostel)

I awoke to first light about 6:30 AM, and initially thought I heard rain on my tent, so I took out my earplugs to listen more. It was just a light drizzle, but the surrounding air was thick with that familiar morning fog. Owls were hooting all night somewhere nearby, but for now, they were silent. I hit the trail about 8AM and had a nice hike with Shutter across Pearisburg Mountain ridgeline all morning. We both had a similar pace, and we stopped at several overlooks with great views and took a lot of pictures. At one of the southeastern facing overlooks, we were standing on an escarpment formed by the erosion of yet another breached anticline where the resistant Silurian Clinch Sandstone is the major ridge forming unit. The cliffs formed here were remnants of the breached (or eroded) southeastern limb of the Bane Anticline. As the fog cleared, we soon had a fantastic view of the Wilburn and New River Valleys, looking south toward Staffordsville. We remained at the overlook with good visibility for about 10 minutes when suddenly the low clouds moved in to completely shroud our view in another thick blanket of fog.

We began our descent down into Pearisburg, and about halfway down the mountain, the rain started again. It was a light, steady rain and was quite welcoming on this warm morning. We got to the parking lot at the bottom by 1 PM and called Angel's Rest Hikers Hostel for a shuttle. While we were waiting under the semi-cover of some large oak trees, I noticed a car parked nearby. Suddenly the front window rolled down and a young man stretched his neck around to say hello. He explained he'd been living out of his car for several days here in the parking lot and was waiting for a government stimulus check to arrive in the mail so that he could continue on his hike. Our ride still had not arrived after 15 minutes, so we began walking down the road in the direction of town. Almost as soon as we started, the shuttle passed us and we flagged him down.

I arrived at Angels Rest about 1:30 with Shutter where we were checked in and shown to an empty bunkhouse for our lodging that night. There was no one else in the bunkhouse, so I took one corner and Shutter took the other. I unpacked my gear and showered, then washed clothes and my sleeping bag. Next, we decided to go find some food for lunch, so we followed the blue-blaze trail through the woods to the nearby shopping plaza and found a great little Mexican restaurant. The jalapeno poppers and enchiladas I ordered were outstanding as were the frozen margaritas. On the way back we went to a grocery store to re-supply. For the next five days of food, I spent $75 total. I also bought a few juicy oranges and apples, and even some Foster's lager for later. We followed the blue-blaze trail back through the woods to Angels Rest. Shutter bought a selection of delicious bakery goods for tomorrow's breakfast. Later that afternoon, I made phone calls home, texted some pictures, and rested. For dinner, we ordered several pizzas to be delivered, and then went over to the lounge building to watch some TV while we feasted on ice-cold lagers, soda, and pizza.

Day 50
8.4 miles

Day 51 August 4, 2020 Tuesday
Lane St, Pearisburg, VA (Angel's Rest Hostel) to
Pine Swamp Branch Shelter, VA
I woke up about 6 AM in the bunkhouse at Angels rest, packed up, and was ready to go by seven. I went over for coffee and breakfast in the lounge with Shutter. He was explaining how he took some alternative classes while attending his high school in Florida, one of which was juggling. We poured some more coffee and moved out to the front porch where he went on to tell me how he could juggle just about anything, with the exception, perhaps, of things like chainsaws, swords, and television sets. Broccoli Bob came by to check on us and we talked about his garden for a bit. I went back inside, picked up three green tomatoes, and brought them out and handed them to Shutter. "Can you juggle these?" I asked. It was amazing how well he could juggle, showing Broccoli Bob and me several different techniques that soon had me imagining I was watching an act from Cirque du Soleil. I was ready to hit the trail for the day, but Shutter had decided to stay at Angels Rest for a ZERO day in order to rest up his legs and feet.

Broccoli Bob shuttled me to the trailhead at about 9 AM. It was a very good trail once I passed a bridge and petrochemical plant. That part of the trail, while not scenic, held a keen interest to me. About a year ago, I recall cyber-hiking my way through Google Earth on my home iMac along the very part of the trail I was now hiking. Ascending Peters Mountain, I met Groundhog, a retired agronomist, or soil scientist, from

Virginia Tech who does trail maintenance work as a volunteer. He then explained how it was that he acquired the esteemed title, "Groundhog." Apparently when he first started his volunteer work as a trail worker, the crew was assigned to a very rocky and dry part of the trail, with very little dirt to be found for filling in the cracks and spaces between the rocks being set. Evidently, he was always able to find plenty, despite the dearth, of dirt and gravel, thus earning him this much sought-after trail name. We talked for about 15 minutes about the general geology and agronomy of this area; it was very satisfying to chat with someone who really understood a bit about geology.

I continued following the white blazes until noon, when I finally arrived at a boulder-strewn bald on the ridge top just past Rice Field shelter. The views were of course amazing, and their breathtaking beauty mesmerized me. While I was enjoying some culinary delights, three hikers from Indiana showed up, sat down, and joined me. We talked until about 1 PM and then I left for points north. It was a very long afternoon. I followed the ridgeline over some incredible structural geology. Clinch Sandstone and orthoquartzite beds were dipping to the southeast along the ridge with nearly continuous exposures.

At 19.6 miles, today was my longest day yet, but I was still eager to break that 20-mile mark! I arrived at Pine Spring about 7 PM. I got water and washed up in a stream, then came back to the shelter, which was posted with a sign saying "Closed Due to COVID-19." Rain started coming down hard, lasting only 15 to 20 minutes. I made dinner under cover, looked for a suitable tree to hang my food bag, and finally set up my tent in a flat clearing nearby. By 9 PM, I crawled into my tent to begin reviewing maps and planning for the next day.

Day 51
19.6 miles

Day 52 August 5, 2020 Wednesday
Pine Swamp Branch Shelter, VA to War Spur Shelter, VA

I woke to gentle raindrops falling on my tent. As it turned out it was merely dew falling from the trees, and I emerged from my Altaplex to a beautiful morning, sunny and 62. I made my usual coffee and oatmeal and was in on trail by 7:45. Later I had to stop for water and biobreak, only to find the much-awaited privy locked due to NFS COVID restrictions. Of all the shelter areas I visited, there were only a few with padlocked privies, and this generally resulted in a feverishly hurried cat-hole digging frenzy.

After the first two hours of gentle terrain, I took a short break against a large oak tree. Following this, the next two hours were a bit more invigorating, as the trail led me over the very steep rocky terrain of boulder fields on the front side of thrusted Clinch Sandstone mountains. Here the trail was essentially taking the unknowing hiker across a talus slope resulting from many millennia of erosion of the northeast leading edge of the orthoquartzite beds capping the top of these ridges. This tedious trek took me close to two hours to get through, and by 1 PM, I made it to Wind Rock for my mid-day break. I stayed one hour to rest my sore feet, eat lunch, and review some maps.

I finally left my lunch spot by 2 PM for my afternoon hike. I was considering another long day, but my left foot was still really hurting. As I was hobbling down the mountain, I encountered my third timber rattlesnake stretched out along the left side of the path. The serpent detected me before I could even see him, as he alerted me with his telltale rattle. I instinctively froze in my tracks—he then slowly slithered away, disappearing into the high weeds and witch hazel underbrush. His rattling continued until he was well hidden again. He meant me no harm and we hastily parted our ways.

I decided to end my day early after just 12 miles, finishing at 4:30 at War Spur Shelter. This was another good call! It was a very nice shelter area, and the privy was thankfully not locked. There were multiple spacious tent sites all around, good water right down the trail about 50 yards, and even a nice bench right next to where I set up my tent. I washed up and filtered two liters of water from the stream and made an early dinner. Because I arrived early, I had more time to read, rest, and relax for a change. I was still worried about the left foot; the area around my metatarsals had become increasingly swollen and painful in the last couple of weeks. Sometimes they would keep me up at night. I had been massaging both feet every night and it seemed to help.

From here, it was only 32 miles to Dragon's tooth, the first destination of Virginia's "Triple Crown." Alone again at tonight's camp, I hadn't seen any other thru-hikers today, just three day-hikers in the morning. There was some discussion about Hurricane Isaias having lost its potency so weather should be better than earlier predicted. The day remained mostly sunny until 5 PM when there came a brief shower. Before sunset, the sky cleared and I could see stars through the thin canopy most of the night.

Day 52
12.7 miles

Chapter Nine
Virginia's Triple Crown

Day 53 August 6, 2020 Thursday
War Spur Shelter, VA to Niday Shelter, VA

    I woke up at 5:30, packed up in early light, made coffee and oatmeal, went down to the stream to get water, and was on trail by 7 AM. There were very nice trails all day. Coming through Johns Creek Valley, I passed through several pastures and then made my way through the mostly hardwood forested uplands of Rocky Gap and Kelly Knob following another Clinch/Tuscarora ridgeline. From there, the trail led me back down again into the scenic Sinking Creek Valley. I pushed on, climbing my way uphill to a breezy and scenic lunch spot on top of Sinking Creek Mountain, where I had more wonderful views to the northwest. It was a very long uphill, but I made it by 1 PM, having logged about 10 miles since leaving my camp this morning. On the way up the mountain, I strolled past the famous 300-year-old Keffer oak tree. I have to say, I've seen much larger and more impressive oak trees, but apparently this is one of the oldest on the AT.

    The hike to the top began in Sinking Creek Valley, locally underlain by Cambrian Elbrook limestone and dolomite. Because of the limestone bedrock, the valley exhibits karst topography, meaning it has many sinkholes and underground caverns. Karst topography occurs when the limestone bedrock chemically weathers in the presence of slightly acidic groundwater. As rain falls, CO_2 is absorbed by each droplet forming a mild solution of carbonic acid, which then reacts with limestone, dissolving some of the rock especially along faults and fractures. This results in the formation of caverns and subterranean drainage. Although the effect of subterranean drainage on Sinking Creek is much more conspicuous to the west as it nears the New River Valley, many of the rocks beneath this part of the valley are similar to those further west where the stream disappears during seasonal low water levels. As I continued up the steep northwest side, I first passed through a succession of shaley, well-graded pathways eventually yielding to the beginnings of the purplish-colored Upper Ordovician shales and sandstones of the Juniata Formation. Then, the large beige to white boulders eroded from the ridge-forming Tuscarora Sandstone were seen more frequently. Finally, the trail emerged over several boulder and cliff scrambles composed of the seemingly ubiquitous Tuscarora. Once I arrived at the summit, there was generally a level ridgeline path along the crest of Big Sinking Mountain.

    I was enjoying a relaxing lunch break at the breezy summit when I began hearing the unmistakable growl of thunder off in the distance. Soon, there was a lot of rumbling but still just a drizzle of rain, continuing for the remainder of the afternoon. Although the ridgeline trail was mostly straight and flat, it was occasionally very rocky with boulder scrambles over the large, heavily jointed and fractured Tuscarora orthoquartzites. Shortly after passing the blue-blaze trail intersection for Sarver Valley Shelter, the trail eventually became a cliff walk along the very steeply southeastward dipping beds of in situ Tuscarora. This part of the trail was treacherous at times, with abundant valley views to the southeast along most of the nearly two miles of this knife-edge trek. Southeast of this ridgeline is Craig Creek Valley, and then Brush Mountain. Beyond Brush Mountain is a long parallel valley occupied by North Fork Roanoke River, which flows southwest toward Blacksburg. The Catawba River, which flows northeast, can be seen from upcoming MacAfee Knob and Tinker Cliffs.

I encountered a fair number of other backpackers today including two Kentucky guys at the Laurel shelter who were both physical therapists right out of school. One of them was complaining about all the blisters on his feet. I helped him with some Leuko tape applied over top of a Band-Aid covering a large painful blister on his heal. He seemed euphoric over just how well this worked showing me with his boot on how he was now able to walk pain free. It also made me feel good to know my near non-existent medical acumen had actually helped alleviate someone's pain and suffering! I later encountered two young ladies just 90 miles from completing their thru-hike for 2020. Scooby and Boogie said they would finish in four months and four days' time by avoiding towns and doing about 18 miles a day average. One of them, maybe Scooby (I really can't recall), described her foot pain and it sounded a lot like how my feet felt! Later, I encountered two West Virginia women on the steep back limb trail, explaining that they spotted an enormous timber rattlesnake on the trail earlier. I continued past them with cautious optimism hoping I might see another, but in a way hoping to avoid the rattler. There turned out to be no snake ahead and I admit I was secretly relieved.

I made it to camp by 5:30 PM, already there were four others: two women in the shelter and another couple, Jay and Christie, in a tent. Jay was a professor near DC, and Christie worked for the government having had several overseas stints. I set up near them due to lack of flat campsites elsewhere. We all made dinner together, and I had interesting chats with all four hikers. I did not anticipate that I would see any of them again due to differing itineraries and planned paces.

Day 53
18.4 miles

Day 54 August 7, 2020 Friday
Niday Shelter, VA to Lost Spectacles Gap, VA

I woke at 5:30 AM, broke down camp, made hot coffee and oatmeal, and was on trail by 7 AM. This morning was overcast, cool, and foggy. I made it to the Audie Murphy memorial monument on Brush Mountain by 9:40 AM. Audie Murphy was the most highly decorated American war hero of WWII, and having died in a tragic plane crash on this mountain in 1971, he is forever memorialized with a bronze plaque and stone monument at the top. There was a bench facing the stone monument where I solemnly sat and had my second breakfast for the next 15 minutes.

I walked back down to the trail, which was an old forest service road, and the high grass quickly made my shoes soaking wet. I eventually arrived at Trout Creek and refilled my water bottles. While there, I met a father and son going south, heading up the trail towards Pickle Creek Shelter. I stopped at a moss-covered, old campsite at noon for lunch. Literally, the moss was blanketed everywhere like a plush green carpet, and I took off my shoes and socks, sitting restfully against a large oak for a much-needed lunch break. While attempting to close my eyes for a short rest, I began noticing a few curious carpenter ants appearing on my feet and ankles. I had been aware of their presence when I sat down on the moss carpet but largely ignored them since ants were pretty much ubiquitous every day. After a few minutes however, I noticed that they were crawling all over my legs and I could soon feel them swarming up my back. I quickly stood up while shaking off the last of them, making the decision to just move on. Soon, I was back on the trail and heading towards Dragon's Tooth, sadly unrefreshed.

I remembered the night before, while reviewing today's map that the trail around Cove Mountain on the way to Dragon's Tooth looked like the number "2." I have since seen all sorts of topographic and geologic maps, including various satellite imagery, and the number "2" pattern is clearly conspicuous on all of them—regardless of scale. The reason for this pattern is best answered by a detailed explanation of the structural geology, recently remapped, described and published in 2019 by Philip S. Prince, of the Virginia Division of Geology and Mineral Resources. Simply put, the odd pattern is where the Clinch/Tuscarora caps the ridgeline, and the shape is the result of the folding and thrusting of several fault blocks during the Alleghanian orogeny some 300 million years ago.

The trail up to and along Cove Mountain was difficult, and it took all afternoon to get to Dragon's Tooth. There were a lot of rock steps and parts of the trail requiring slow, careful climbs up and down the massive sandstone boulders. By 4 PM, I arrived at this spectacle, and my first thought peering up at the nearly vertical sandstone and orthoquartzites beds was, "This is the Seneca Rocks of Virginia." In West Virginia, Seneca Rocks is the remnant vertical northwest limb of Wills Mountain

Anticline, where the Tuscarora is the ridge-forming unit, noticeably visible from miles away. The Tuscarora, or Clinch—the name more commonly used toward southwestern Virginia—is the same ridge-forming unit found here at Dragon's Tooth.

While milling around the ridgeline, I met several young day- and section-hikers and we talked about the amazing rock formations and the trail conditions leading us here. I only stayed for about 30 minutes, then headed north down the mountain. It was a very enjoyable but difficult descent. Coming off the steep mountain, there were rebar ladders anchored to the vertical rock face, cliff walks, and house-size boulders to negotiate. All of this was a lot of fun if you were not exhausted like I was. I really could've used that noon rest time. Darn those ants!

I eventually arrived at Lost Spectacles Gap by 5:30 and set up my tent in a cozy flat spot next to a cluster of rhododendrons. The gap here was peculiarly named for Tom Campbell, one of the early twentieth century trail builders on this intriguing mountain and where he allegedly lost his spectacles during one of his outings. While I was setting up and making dinner, there was a constant flow of hikers returning from their day trip to Dragon's Tooth. One family stopped to say hello and we talked for about 15 minutes. They all seemed very interested in hearing about my thru-hike, questioning me about my gear, plans, and past adventures, which I cheerfully answered. I was in my tent by 7 PM, and by 7:30 very heavy rains began, persisting through much of the night.

Day 54
15.7 miles

Day 55 August 8, 2020 Saturday
Lost Spectacles Gap, VA to Tinker Cliffs, VA

I woke up at 5:30 AM to the calls of a couple of bobwhites. I made cold coffee and ate a Clif bar and was on trail by 6:50 AM. I had terrific views from the many overlooks all the way down to a creek. The morning was foggy and cool, but I could see out over the clouds, which were blanketing the valleys. Although very rocky and difficult to navigate in places, the trails were mostly smooth and well graded all morning. I stopped at the creek near a road at the bottom and started up the mountain toward McAfee Knob—a 9-mile hike. When I got to the parking lot for McAfee Knob, it was totally full. Incoming cars were literally waiting for day-hikers to return to their cars in order to free up spaces. On my way up the four miles from the parking lot to the overlook, I encountered more people in two hours than I had seen in the past month. Everyone seemed very friendly, and I had nice chats with many of the day-hikers.

I reached McAfee Knob about 12:30, where once again, all around me on this ridge top appeared to be the familiar Silurian Clinch, or Tuscarora orthoquartzite. However, recent field mapping suggests that the Tuscarora is not mappable here and rather, these rock units are mapped together as Devonian-Silurian sandstones. Beds here are locally dipping back to the south, as part of a thrust sheet from Alleghanian time generally being shoved from the southeast to the northwest. Looking into the Catawba Valley, I was peering into older, Ordovician-age shale, siltstone, and limestone beneath a thin veneer of forest and agriculture. That same thrust, known as the Pulaski thrust fault, brings basal Cambrian Elbrook Dolomite in contact with much younger Mississippian-age sandstone and shale formations 2 1/2 miles northwest of McAfee, near the base of North Mountain. I walked out to the cliffs to gaze at the incredible view, and then found a nice spot near the overlook to finally sit down and enjoy my lunch. While eating, I once again ran into the family that I met at my campsite last night at Lost Spectacles Gap. Last evening, they had been very excited to hear about my thru-hike, and today they brought with them that same enthusiasm. We were glad to see each other again and talked for a few minutes. They were very thoughtful and compassionate and wanted to do something for me, so they offered me several energy bars and an ice-cold Gatorade. Wow, more trail magic! I really appreciated that more than they knew. I thanked them profusely and after a short chat we said goodbye. Later, a very kind lady agreed to take my picture standing on the legendary overhang rock. When she came over to return my iPhone, I thanked her and asked where she was from. She replied "Crafton, near Pittsburgh, Pennsylvania." What a small world! We talked for a while about Pittsburgh, and how much we both enjoyed living there.

After MacAfee, I hiked one mile to a spring for some water. While there, I met a friendly couple having their lunch. They met at a local bible college and were just hiking together for the weekend. After a brief chat, I walked back to the shelter, had more lunch, and took a very short catnap with my shoes and socks off. I popped an ibuprofen and gobbled down another Clif bar feeling totally recharged for my hike to Tinker Cliffs. Within the first mile, I caught up to three young people who I had met while having lunch earlier. I didn't get their names, but there was one very tall young man and two talkative friendly young ladies—all college age—and we hiked together for about an hour until they decided to stop for a break. From this point, the last mile or so up to Tinker Cliffs was brutal. The trail leading to the cliffs and the cliffs themselves consist of Silurian Tuscarora once again. This formation is mainly a conglomeratic orthoquartzite, meaning simply, a silica cemented sandstone with grain size in excess of two mm. Very tough rock indeed.

The trail was very steep and rugged, with many false summits as was so common on the AT. I passed a very distraught looking young couple who clearly were both exhausted. The poor girl's face was bright red with bulging vacant eyes, just staring into space, with her male companion looking away not knowing what to say or do. I really felt bad for them, especially knowing that there was no easy way out from here. This was a very difficult ascent. For the next ten minutes or so, I just kept wondering about how on Earth they would ever get back to their car for the night. I eventually made it to the cliffs and they were truly spectacular. I could see all the way back to MacAfee Knob, and down into the Catawba Valley, a beautiful pastoral valley with just a few homes and farms scattered about. I hiked for another quarter mile along the tough and resistant Tuscarora cliff walk and decided to camp here for the night. I knew camping was not allowed here but I too was exhausted and needed to stop for the day. I had only hiked 16.7 miles today, but the terrain was very challenging, making it seem much longer. I picked a flat, pine-straw-covered spot that was not only perfect for camping, but the west facing views at the overlook were perfect for spending the next couple of hours watching the sun go down over the horizon.

When I had first arrived at the cliffs there were two young deer that seemed to be following me. Later, when I got to my campsite I noticed them again and they seemed unusually tame. I set up camp, unpacked, and organized everything inside my tent. I then set out to make my usual dinner of ramen noodles on the cliffs. Off in the distance, a solitary dark cloud appeared to be dropping rain on the far ridge. As I watched this for a while, I could see that it was slowly coming my way across the valley. The sun was shining on me the whole time, as it was low in the sky, nearing sunset. It had been a hot, sticky day and I could still feel the humidity in the air. Soon I heard the rain hitting the leaves on the trees in front of me and then suddenly the refreshingly cool rain was falling all around me. It felt wonderful. This lasted no more than 10 minutes and the sun remained out throughout the entire little shower. I continued watching the sunset, as it slowly dropped towards the horizon, illuminating the clouds in a changing array of subtle pinks, yellows, and oranges.

Later that night while I was in my tent, I heard clomping behind, and then in front of the tent. I slowly peered out into the darkness shining my headlamp toward the sound, and I saw two glowing eyes staring back at me from about 20 feet away. I held the light steady as the eyes remained motionless in the sea of black for the next ten seconds. I waited for some indication of identity or intent when suddenly they began bobbing up and down. "Ahh," I concluded to myself. Apparently one of the deer from earlier was still scouting around my camp. Just curious, I guess!

Tomorrow's resupply in Daleville was just 10 miles away. I planned to arrive there by noon and then either stop at a motel or hostel for the night—or perhaps just continue north after restocking my food bag. Either way, my decision would have to wait until morning.

Day 55
16.7 miles

Day 56 August 9, 2020 Sunday
Tinker Cliffs, VA to Fullhardt Knob Shelter, VA

I woke at 5:30, broke camp, and was on trail by 7 AM. This morning was cool and foggy, but I could see the sunrise through the trees and fog to my east. The views were literally indescribable this morning, with all the clouds nestled in the Catawba Valley below. I wanted to stay longer and just watch the morning unfold, but I needed to move on. I followed the cliff trail around to the descent trail, which led me to a shelter where I was able to collect water and use the privy. There I met three young men who had been camping down near the stream and we chatted for a while. Later as I was hiking alone further down the trail, the three caught up to me and we all hiked together for the next hour or so. One of them was a financial advisor from Charlotte, North Carolina. As we hiked, our discussion led to the effect of COVID on our economy and why the market still had not responded to the dramatic and historic 33% drop in GDP. His opinion was that we were inevitably in for a major correction sooner or later.

Much of the trail followed mountain streams, leading to very pleasant hiking. Up ahead, I noticed a single tent next to the stream and as I passed someone called out, "Hey Old Oriskany!" I stopped in my tracks and turned around. It was Honey Mustard, my chef friend from Colorado. The last time I had seen him was my final night in Damascus, when he and his chef buddies all made us dinner on a charcoal grill in the backyard of the Broken Fiddle. He said he was going to hike into Roanoke today, and then take a bus to Maine tomorrow. We talked for a little while longer, wished each other good luck, and I was back on the trail again.

I continued to Hay Rock, the 1/3 point for NOBO hikers. I met an older couple along the way and we hiked together for the last mile or so. We talked about my two-month-long adventure so far, and my plans for the rest of my journey. They kindly offered me a ride to a shopping center on Route 220, about a mile north of the AT intersection with the highway. I hopped into the back bed of the pickup truck and was soon dropped off at the local shopping area. I looked around and noticed there were several restaurants to choose from: fast food joints, a Chinese take-out, and an outdoor Barbeque place. Without a second thought, I marched straight towards the Three Little Pigs Barbeque, where I sat at an outside table and ordered a rack of ribs and a local IPA. I was outside and far enough from everyone else that I didn't worry about offending anyone with my thru-hiker aroma. My waiter was friendly and quite interested in hearing all about my AT adventures and said that he wanted to hike the entire trail one day. After stuffing myself with barbecue, I wandered over to Kroger for my resupply. I have long known that it is far better to food shop on a full stomach than an empty one. However, perhaps as result of knowing this, and in an attempt to overcompensate for my utterly satisfied state, I found I bought far too much food—probably enough for seven or eight days.

After re-packing everything, I threw my now overweight pack over my shoulder, and plodded over to a gas station where I noticed a gentleman filling up his pickup truck with gas. I asked him if he lived near Troutville and could give me a lift. To my surprise he said that he actually lives on the AT and would be happy to give me a ride. This was more than I expected! He also explained that his family used to own one hundred acres on the very mountain I would be camped out on tonight, Fullhardt Mountain. He later dropped me off at the trailhead around 4 PM, and I hiked up the mountain to Fullhardt shelter. After about 30 minutes of a fairly arduous climb over Cambrian aged-sandstones and shale, I was almost to the summit. The top of the mountain was now underlain by a more resistant quartzite of the Antietam Formation. I had officially left the Valley and Ridge and was now back into the Blue Ridge province.

Approaching the top, I began to hear the sound of children laughing and dogs barking. When I finally arrived, there was a family sitting at the picnic table with three kids and two dogs, and they were just finishing their lunch. All of them were very convivial, and after 20 minutes or so of lively chitchat, we all said goodbye and the jolly picnickers headed back down the mountain.

Fullhardt Shelter is on a mountaintop and has no access to a spring. For water, there is a collection system coming off the roof, which fills a cistern underground. About 100 feet behind and downhill is a spigot that delivers water from the cistern. There is also an old foundation between the shelter and the privy from some type of fire or radio tower long since dismantled and removed. Upon inspecting the privy, I found it thankfully unlocked, and home to some of the largest spiders I have ever seen on the trail. Later, I set up my tent on a rare flat spot way down the path and slept with a pleasant forest breeze wafting through my screen door all night.

Day 56
15.5 miles

Chapter Ten
Breaking New Records

Day 57 August 10, 2020 Monday
Fullhardt Knob Shelter, VA to Cove Mt. Shelter, VA

I awoke at 5:30 AM and looked out the screen on my tent to see pink sky through the trees. It was cool and a bit foggy, almost like I was in a cloud. I was anxious for today because I would finally reach the Blue Ridge Parkway at Blackhorse Gap. For the next week, my path would parallel or cross this famous roadway many times. When I arrived at the Blue Ridge Parkway around 10:30, there were numerous roadside rest areas with beautiful vistas of the surrounding pastoral valleys. I stopped at one and talked with a family about the loss of the chestnut trees in the past century. We were also gazing out the west trying to identify small towns, highways, and rivers we could see in the distance. I used my paper map for reference in identifying all the questionable features in the horizon, and they seemed surprised by this ability.

I had a long morning hike and made it to Bobblets Gap for lunch about 12:30, after a grueling 13-mile march. Because I hadn't run into many other hikers, I was breaking spider webs continuously as I slogged through the wet, muddy trail for those 5 ½ hours. I could always tell if there was anyone somewhere up ahead of me by the number of spider webs breaking across my face. It was always a relief when I would finally pass a SOBO hiker because now all of the webs would be gone—for a while at least! It was a relatively easy day today, no blisters yet for which I was thankful. I also ate a lot of energy bars, one every hour or so all morning, for continued energy.

Soon I was off again heading north on the trail, and by 5 PM, I got to Cove Mountain Shelter. When I arrived, there was a young couple with their daughter Amelia, a very talkative nine-year-old young lady. Her parents were Ann and Jeremy from Durham, North Carolina, both very friendly, and we ate dinner together around the picnic table. I shared stories about my adventures over the last two months and did my very best to answer all the questions from young Amelia. We all seemed to enjoy each other's company, and the four of us continued chatting well after dinner. I retired to my tent by 7:45 feeling quite satisfied with my 20 miles today. I had finally broken the 20-mile mark!

Day 57
20.0 miles

Day 58 August 11, 2020 Tuesday
Cove Mt. Shelter, VA to Thunder Hill Shelter, VA

I woke at 5:30 AM, broke camp, and had hot coffee and oatmeal while watching the sunrise through the trees. I was on the trail by 6:45, and after a pleasant morning hike, arrived at the first shelter at Bryant Ridge by 10 AM for my second breakfast. I made it to the second at Cornelius Creek by 12:30 for lunch—so far 12 miles for the day; another great start! Lots of uphill climbs culminating in satisfying ridge walks with a cooling breeze beginning near summit and following me up and over each mountain I climbed. Most of the rocks along the trail were Cambrian quartzites, Proterozoic gneisses, and other assorted metamorphics. After filling water bottles at Cornelius Creek, I rested against my backpack up against one of the timber pillars inside the shelter. While eating my lunch, a man walked in from the Blue Ridge Parkway with his lunch as well, looked at me and said, "I was hoping I would find you here." I wasn't sure what to make of the comment and said hello and we talked for a bit. It turned out he was a retired religion professor at Liberty University in Lynchburg now pastoring his own church. He offered me an apple and a peach, as well as pretty much anything else in his lunch bag. I gratefully accepted the fruit and thanked him saying I hadn't eaten fresh fruit in over a week. Apparently coming here every few days was his mission and way of helping out hikers on the AT. I appreciated his benevolence and kindness, and we talked for about a half an hour before he was on his way again. After he left, I took a short 10-minute rest and was once again recharged for my afternoon trek.

    I was back on the trail by 2:15 and hiked another five miles up to Apple Orchard Mountain—well within the Virginia Blue Ridge Complex, a Precambrian-age assortment of various igneous rocks such as granite and diorite, originally emplaced about one billion years ago. The trail began very steep and rocky, but soon became much better graded and easier than I expected. By 5 PM, I arrived at Thunder Hill, where I found a nice tent spot, and went looking for water. The water source here was a small, circular stone well, with just enough water in the bottom that I was able to carefully scoop out cup by cup to fill my water bottles. In addition, I collected some more water to take back over to my tent area, where I could wash off the sweat and grime from today's hike. In this humidity I was sweating more today than I ever can remember, and the gnats were becoming intolerable. I got organized and was preparing dinner when another hiker arrived. Walton, a SOBO section-hiker from Richmond had been working on a farm for the past several years, which was part of a facility providing care and training for handicapped persons. We had dinner together and talked all about the high points from where we had just hiked, either NOBO or SOBO. I found this very exciting and helpful to hear about what was in store for me in the weeks ahead.

    Behind the shelter there was a bear box, so no need to hang the food bag tonight. This was a luxury and always saved me time by not having to search for a suitable tree or attempting to throw my rock bag over a high limb. Later, I headed back to my tent for the night, while Walton waded through the high grass searching for a suitable location for his sleeping bivy.

Day 58
17.2 miles

Day 59 August 12, 2020 Wednesday
Thunder Hill Shelter, VA to Punchbowl Shelter, VA

I awoke at 5:30, broke camp, and had breakfast of hot oatmeal and coffee. I packed up and was ready to go by 7 AM. Walton's clothes had been dragged from the deck of the shelter down into the mud, and were lying there in a heap. Upon further inspection, it looked as if something had been chewing on them. Using a long stick, I carefully placed them back up on the platform, but didn't bother to wake Walton in order to tell him what happened. I headed up the blue-blaze toward the AT and started on my hike for the day.

It was cool and misty but clear skies were present overhead. However, the weather forecast was predicting rain for the next five or six days beginning today. There were lots of easy downhill trails early on, and I made it to Highcock Knob by 9:15—six miles in—for my second breakfast. At the 13-mile point for the day, I arrived at the next shelter after crossing over Matts Creek. This was a brisk mountain stream slowly cutting its way through metasedimentary rocks of the Cambrian Chilhowie Group. Along its path were numerous gushing waterfalls punctuated by deep, clear pools. I quickly prepared one of my cheese and pepperoni wraps and walked barefooted over to the stream submerging myself—clothes and all—into the refreshingly cool, sparkling water. As I relaxed in my natural Jacuzzi, I watched in amazement at dozens of small minnows nibbling away at, not only the fallen crumbs from my wrap, but the dead skin of my feet as well. This was a kind of minnow metatarsal manicure!

All morning I was thinking a lot about a speaker opportunity that I turned down this week. Linda, a fellow geologist and friend from the Houston Geological Society, had invited me to participate with several others in a Zoom chat to discuss a West Virginia field trip held a few years back. The live chat was to be broadcasted to an audience of geologists in the Houston area. I didn't know where I would be or whether I would have cell service by Friday at 11 AM, and so regrettably declined the invitation. Was there some way I could have done this from the trail? I knew the answer was probably "no," primarily due to poor cell service, but it would have been fun to report to a large audience of geologists from out here on the AT.

After lunch, I followed the trail along Matts Creek to the James River. From there, the white blazers led me along the James to a footbridge across the river. I somehow missed the bridge and continued another 1,000 feet or so to a railroad trestle. Realizing I went too far, I turned around and backtracked to the footbridge. As I was climbing the steps up to it, I shouted hello to a young couple in their bathing suits, getting ready to go tubing in the river. It looked like the perfect activity for a hot afternoon, but my goals were set for other, far less relaxing activities today. I crossed the river and continued towards the next shelter where I thought I might stay for the evening. When I arrived, it was still early afternoon, so I continued up to Little Rocky Row. Once on top, I figured I could find a nice tent site for the night, maybe even with a view.

The climb to the top of Little Rocky Row was brutal, with parts of the trail regularly steepened by resistant quartzites and other metasedimentary strata outcropping everywhere. I was tired, hot, and sweating profusely as I climbed this steep ascent, and the gnats were intolerable. One or two of them would be endlessly flying around my face, and about every seven seconds or so, would dive-bomb my eyeball. This continued indefinitely. Despite the annoyance, I finally made it to the top to see some nice storms approaching, revealing to me that I was about to get drenched. I found a perfect campsite just before the overlook, but wasn't ready yet to call it for the day, and I was sure I would see more tent spots a little farther down the trail. I was so wrong! Amidst the far away flashes of lightning and rolling thunder, I climbed up to Big Rocky Row. Again, the heat and the gnats were driving me crazy. Suddenly, the wind picked up, and that far-away lightning and thunder was now happening all around me. Next, the rain began falling, at first lightly as a mist carried by the wind, then changing quickly into a deluge. My mind raced to Mt. Doom in Middle Earth. I continued up the precipitous trail, having to make frequent scrambles using both hands to pull me up the rock cliffs while the horizontal rain pelted me—flashes of lightning all around.

My plan today was to find a campsite somewhere along a ridgeline, but there were none. After getting drenched and cooled by rain, I continued down Rocky Row along very nice flat and fast trails. It was extremely pleasant hiking, especially compared to what I had to go through to get here. By now the rain had ended, and there were some signposts along the way indicating Bluff Mountain was ahead. The trail was nicely graded all the way to the top enabling me to maintain a faster than usual pace. However, it was getting late. I was exhausted, but I had to continue, looking for anywhere I could stop to set up camp for the evening. The narrow and rocky ridge top to my right and the steep drop off to my left afforded no opportunities for a flat campsite. I plodded along for another 1.5 miles, finally arriving at Punchbowl Shelter by 8 PM. My tank was on empty, and I was beyond exhausted.

As I lumbered down the blue-blaze trail toward camp, I noticed the unmistakable aroma of burning firewood. My spirits suddenly lifted as I saw a figure sitting by a blazing campfire up ahead. Inside my head I shouted, "Somebody made a fire!" Nick was from Suffolk Virginia, a ship worker at the navy shipyards, and was out solo backpacking for a few days. I walked over to say hello, wanting to shout hallelujah, and we chatted for a few minutes. I hurriedly pitched my tent, unpacked, and rushed back over to the warm fire pit to prepare dinner and talk to my new friend. The rain had been misting, but now began to pick up to a steady downpour. We both sat under the cover of the shelter by the fire as the rain came down around us. I stayed up longer than normal that evening, but after today's new record of more than 25 miles, I felt I deserved it!

Day 59
25.3 miles

Chapter Eleven
Raining Raccoons and Washed Out on the Priest

Day 60 August 13, 2020 Thursday
Punchbowl Shelter, VA to Cow Camp Gap Shelter, VA

Yesterday was a very long day. I didn't set an alarm, so I slept about one hour later than planned. I was on the trail by 8:15, with mostly downhill until I reached the gap before Rice Mountain followed by a very nicely graded climb to the summit. From there it was all downhill again to Lynchburg Reservoir, a dammed portion of the Pedlar River, which eventually flows into the James. Just before the reservoir, there was a narrow footbridge crossing the Pedlar. As I crossed the footbridge, I noticed a large cicada caught in a spider web, still alive and struggling to escape. I carefully flicked him into a clear, deep pool just below me in the river. The oversized insect landed on the still water with a plop, and I watched, mesmerized, for about 10 minutes as one trout after another rose attempting to grab him. Unfortunately for the fish, the cicada was too large for the mouths of any of them. After passing by the large scenic reservoir, the trail followed a relatively flat section leading north towards Brown's Mountain Creek.

At Brown's Mountain Creek, I came across an old settlement where freed slaves lived after the Civil War. There were lots of old stone foundations and walls hinting at what a bustling little village this must have been at one time. From here, I began my long ascent to Bald Knob. Ahead of me were about four miles of steep trail where I would gain 2,600 feet in elevation to reach the summit. The rain was just a drizzle and would help to keep me cool and keep the gnats at bay. The path to the top was surprisingly less difficult than I expected, but still a very long, hard climb. Later, I had a 90-minute lunch and quick power nap to recharge for what was in store for me in the afternoon. From Bald Knob there were glorious views, however I couldn't use the camera on my phone because everything was so wet. I sat on the massive ledges of Grenville-age gneiss, taking in a great view of fog rising from the valley below, and clouds rising with more rain. It was truly spectacular to watch.

I continued on the ridgeline trail down the other side into Cowcamp Gap, where a sign indicated Cowcamp Gap Shelter was 0.6 miles south on Old Hotel Trail. I followed the blue-blaze trail to the end and arrived at my destination about 5:30. I set up my tent, got water, and cooked up my usual batch of ramen noodles for dinner. I was in my tent and finally dry by 8 o'clock. Just then, I heard two hikers strolling by and continuing up the hill to another tent site area. I came out a few minutes later and went up to say hello. It was a young couple, Mama Bear and

Griswold, from Blacksburg with their dog, doing a section hike for four to five days. I returned to my tent and continued my reading and journaling. Soon after, I heard a lot of scurrying outside and the sound of what I thought might be raccoons. My tent was set up under a scraggly witch hazel tree, and it sounded like an animal was in the tree above me. Suddenly, something hit my tent, slid down the side and scampered away. I can only assume it was one of the young raccoons learning to climb. Tomorrow I would attempt to make it to the Tye River Valley, hopefully finding a shuttle to a hostel for a break. My feet were getting blisters from this never-ending rain, and I was desperately in need of a break to rest and dry out for a while.

Day 60
15.1 miles

Day 61 August 14, 2020 Friday
Cow Camp Gap Shelter, VA to The Priest Shelter, VA

I woke up at 5:30 AM, organized everything inside of my tent, broke down camp, had breakfast in Cow Camp Shelter, packed up, and was on my way via the blue-blaze Old Hotel trail by 7:30. The rain stopped and the weather was pleasant but overcast and cool, about 65°. This morning's climbs consisted of multiple smaller mountains, each with no more than 500 feet of elevation gain. I could generally see light through the trees at the crest before I began my ascent. Many of these peaks were balds with recently mowed meadows on top, such as Cold Mountain, Tar Jacket Ridge, and others. The balds here seemed somehow different from those in North Carolina. These appeared more like farm pastures rather than natural meadows, perhaps also because they had just been mowed. When I arrived at the summit of Cold Mountain, I noticed that I finally had cell service and so I made a couple of quick calls to Josh and Jessie. (This was the morning of that HGS Zoom call I declined to attend, and at least for the time being, it appeared that I might have had cell coverage). We talked very briefly and I moved on. I also attempted to send some pictures home to family, but the poor cell service resulted in an error message reading, "Not delivered."

I ran into one enthusiastic section-hiker in the morning, Bill Spears, or SBill, from Spotsylvania, VA. SBill Was doing the section from Harpers Ferry south to Springer Mountain, Georgia, with plans to complete the north half next spring. He mentioned that last night at Seeley Woodworth Shelter, "...it was just me and Elvis," and I wasn't quite sure what that meant. (This mystery would be revealed the following day). He also explained that he left some first aid supplies as well as a single packet of Spam inside the shelter for anyone who might need such items. I explained to him that all my tuna and protein were gone, and I would gladly take him up on the Spam offer when I arrived there later today.

SBill also had a unique way of carrying his water; each liter bottle was duct taped to one of his trekking poles. I wasn't sure I would like to hike all day with that much weight on my arms. He also shared with me one of his homemade dehydrated fruit roll-ups, which was delicious. I thanked him, we chatted for a little while longer, and then we were on our way. Later I stopped at Seeley Woodworth; no one else was there, and I found that Spam that he had left as trail magic.

The afternoon was much like this morning, generally pleasant, with occasional light rain to keep me cool. I continued my attempts to call for shuttles and resupply but could not get through to anyone. Finally, I decided I would just have to make it to Waynesboro with the food that I had left. Most of the afternoon was pretty gentle terrain, however the last several miles featured some steeper than usual hill climbs. The rocks here were mostly plutonic, meaning they originated as molten magma deep in the Earth's crust, then slowly cooled, in this case about a billion years ago. Some of these rocks were granite, some diorite, but most were part of the Pedlar Formation. The final 0.9 miles to the next shelter seemed overly arduous, keeping my head down locomotive style, while trudging straight up the mountain; it was late afternoon, and I was becoming fatigued. Eventually I arrived at Priest Shelter by 5:20 PM, found a suitable tent site, set up my tent, unpacked, and filtered and refilled water. Later, my son Josh FaceTimed me from Singapore and we talked for about 15 minutes. I got to see my youngest grandson, Braeden, standing for the first time!

It was a good day, but already the temperature was dropping, and I could tell that it would be very cold and damp in my tent that night. The weather forecast called for heavy rain over the next 24 hours with clear skies beginning the following evening. Yay! Around 7:30, a hiker by the name of Turtle showed up at the camp. He was a SOBO section-hiker (beginning at Harpers Ferry) from Asheville North Carolina, where he played saxophone for a local jazz band. We gabbed for a while and by about 8:30 I was in my tent for the night.

Day 61
16.8 miles

Day 62 August 15, 2020 Saturday
The Priest Shelter, VA to Tye River, VA 56 (pick-up for Stanimal's)
　　I woke up about 6 AM to a wet tent floor. It had rained hard all night, and now the air was chilly and damp, with rain still coming down. Yesterday, I had set up the tent on a nice flat spot, with all of the leaves raked away to expose compacted black organic forest soil. After many hours of intense downpour, this compacted surface quickly turned to a soupy mud mix. Both tent and Tyvek were now filthy from the splashed-up mud. The horizontal vents surrounding three sides were

also splattered with the black mud. The source of water on the floor proved to be my fault, improperly setting up the tent so that the normally down-sloping vents were up sloping due to my ditty bags depressing the far wall of the bathtub liner. (Note to self, don't do that again!) I packed up all my (mostly wet) gear in the rain and quickly moved everything under the cover of the shelter. I finally sat down on the adjoining deck and made hot coffee and oatmeal while talking with Turtle, who had wisely set his tent up inside and under cover. Following our discussion, I decided to call Adam "Stanimal" Stanley, and arrange for a shuttle pick up to his hostel to spend the rest of the day and that night. I hiked out by 8 AM past the prominent Priest lookouts, which were unfortunately all fogged in. Despite the low visibility, it was a very enjoyable downhill hike, with the weather improving the closer I got to the Tye River valley. The previous night was wet and cold, but I would soon be on my way to Stanimal's to dry out.

I got to the parking lot around 10:30 and had my second breakfast while I waited for the shuttle. It was a full 10º warmer down here, and temps were climbing by the hour. Cora arrived at 11 AM driving a new model minivan—not what I was expecting, I guess. I sat in the back as a COVID precaution, and we had a lively conversation on our way back to the hostel. The drive back was about one hour, to my surprise, and we arrived at Stanimal's around noon. I was given the tour and assigned a bed for the night. I unpacked, dried out my boots, sleeping bag, and pad, and then hosed down my tent and Tyvek ground cloth in the back yard. I then hung everything out in the now misty rain, in a feeble attempt to dry. Later I washed clothes, showered, and then walked across the highway to a Dollar General store for a resupply.

I met several hikers who were also staying there: Helo, Elvis (mystery solved), Gary, Water Queen, and a few others. On staff were: Roman, a bio-geography major at George Washington University who was originally from New Mexico and had worked in the Shenandoahs for the last several summers; Glen, a business manager who apparently ran the day-to-day operations here in Waynesboro; and Rumble, a 6-foot 8-inch young man who worked on staff and ran many of the shuttles as well. Later that evening, I joined Elvis and Helo for dinner at a Mexican restaurant just a few blocks away where we sat at a table in the back of a Mexican convenience store. Despite this unusual setting for a restaurant, the food was authentic Mexican. I had the Fajitas Supremas platter, and it was excellent. The rain never stopped all day, including to and from our dinner excursion. After returning, several of us talked for bit longer, and I was in my bunk by 10 PM.

Day 62
4.8 miles

Day 63 August 16, 2020 Sunday
Tye River, VA 56 (drop off from Stanimal's) to Cedar Cliffs, VA

I woke up at 6 AM, came downstairs for coffee and oatmeal as well as several other goodies put out by the kitchen staff. I talked for a while with Roman and a few others then went into the office with Glen to settle up my bill for the stay. My shuttle that morning left at 8 AM, and I was joined by two other hikers; Pollyanna, a 68-year-old woman almost finished hiking the AT as a section-hiker, with seriously torn meniscus; and Gary, a slack packer out for two or three days. Despite her injury, Pollyanna seemed determined to finish this year. Our driver was Rumble, a past river guide who once managed one of the largest sailboat marinas on the East Coast. He explained to us how so many people get tired of owning sailboats, and then literally sell them for pennies on the dollar. He recently purchased a beautiful lightly used sailboat for just $212 from the previous owner who wanted nothing more than to be rid of what he considered a ball and chain around his neck.

Rumble dropped us off at the same parking lot where I was picked up the day before, right off Route 56 near the Tye River. While we were getting our gear ready for our departure onto the trail, I met Slim Jim, an ATC Ridge Runner and general trail maintenance person, who I would later encounter again on the trail. Gary and I decided to hike together and were on the trail by 10:30. Our six-mile climb up 3,000 vertical feet to the top of Three Ridges Mountain took us just over two hours that morning. We were both very proud of our pace and performance. I told Gary I needed to have a short break, but he continued on, explaining that he had a birthday party to get to later that day, so he needed to keep moving.

The rest of the day was much easier than our morning marathon. I didn't stop at any shelters, and really took no other breaks, but I finally had a late lunch around 4 PM at a rest-area picnic-table next to the Blue Ridge Parkway (BRP). While I was on break, several motorcycles appeared to be attempting to set new speed records for the BRP. (Where's a Park Ranger when you need one?) Following that, the last four miles of the day proved to be pretty tough. The profile in my map book made this section look flat, but the reality was constantly changing steep ups and downs the whole way. Worse, most of the trail here involved rock-hopping from one Catoctin metabasalt boulder to another. These rocks originated as Neoproterozoic to Cambrian-age lava flows, which occurred as result of the breaking up of the supercontinent Rodinia. Commonly known as greenstone, these silica-poor, chlorite-rich basalt flows were later metamorphosed. Even when dry, these rocks can be slippery, making travel slow going the entire way. These slick, smooth rocks were further dampened by the frequent springs along the footpath; never drying out due to their northwest exposure. For the next week, I would often be hiking over these same rocks along the highland trails through the Shenandoahs.

By 5:30, I noticed a sign next to the trail with an arrow pointing to the left, reading, "Overlook." I followed the blue blazes to the viewpoint, which opened to a beautiful western exposure. This would be perfect for tonight's sunset. I looked around and noticed that there was a stealth tent site 50 yards or so from the overlook. I set up my tent, took my cook stove and my titanium cook pot along with what I needed to make dinner, and set up on one of the large boulders on the overlook. I found a comfortable place to relax and slurp my noodles while watching the fading pastels of the sunset. It was a beautiful view, and I could even see the glimmering lights from Waynesboro off to the north. Later I FaceTimed my son in Singapore and texted some pictures to all the folks at home. After a satisfying and relaxing end for the day, I retired to my tent by 9 PM.

Day 63
15.0 miles

Chapter Twelve
The Shenandoahs

Day 64 August 17, 2020 Monday
Cedar Cliffs, VA to stealth campsite on old fire road, VA

I woke up at 5:30 and was on trail by 7 AM. There was a beautiful sunrise behind me illuminating the sky with an orange glow. I walked over to the viewpoint for a few minutes to finish coffee and watch the sky go from brilliant orange to pink. I passed by Cedar Cliffs on the trail just past my campsite. By about 7:15 I arrived at Dripping Rock and left a conspicuous note with sticks and pebbles on the ground at the trailhead reading "7:30," for Gary, whom I expected would be parking at this lot for a northbound hike today.

The hike was very pleasurable all morning; I climbed Humpback Mountain, which had lots of well-placed stone steps and great views the whole way. Although I was still in the greenstone, most of the trail was along ridgeline so was not as slick as yesterday. At the top of the mountain, I could see all the way back to Wintergreen Ski Resort. After this summit, it was mostly downhill or flat to Rockfish Gap. I passed many day-hikers, including lots of masks and over-the-top avoidance tactics—very bizarre. I talked to Gary for a few minutes during my lunch break at Stephen Wolf Shelter, and later that morning I met Cobbler along the trail.

After lunch I headed down the mountain and had to cross Mill Creek at a very high water level. It was tricky hopping from boulder to boulder over the high water, but I made it without stepping in. I was feeling like I was actually becoming somewhat of a pro at this rock hopping, but I would be humbled many times over the course of my adventure. As I was filling my bottles I met Unicorn, from Connecticut. She was a section-hiker who I heard mention of by some hikers that I met earlier. We chatted for a while and eventually I moved on. Later, I met a young lady on her first day of the AT who was going south to Springer. She was excited to be starting out, nervous about her late start, and eager to listen to any advice I had to impart on her. Slim Jim was doing trail maintenance with his machete, and around 2 PM I ran into him again, and we stopped to talk for a bit. I had seen him the day before at the parking lot at Tye River where he was painting one of the AT bulletin boards. He told me that he had been an attorney for 20 years and gave it up to work as a ridge runner for the ATC. He also offered for me to give him a call when I got to New Hampshire, where by then he would be doing pre-winter maintenance, in the event that I needed any assistance.

The weather had been beautiful all day, and there was an absolute zero percent chance of rain in the forecast. However, about 3 PM, the skies opened up and drenched me in heavy rain for the next 15 minutes.

This was really no big surprise. I had come to expect these sudden afternoon showers nearly every day since the start of my journey.

I eventually arrived at the entrance to Shenandoah National Park at 5 PM and checked in at the self-serve kiosk. It was nothing more than a box containing forms to fill out so that the Rangers knew who's there and for how long. It had been a very long day and I was ready to set up camp somewhere. After another hour of hiking, I desperately needed to find a stopping point for the day. Continuously scanning the woods to my right and left, I ultimately settled on an abandoned fire road about 100 yards off the trail, which appeared flat and somewhat secluded. I set up my tent, did my evening chores, had dinner, looked for a bear hang for my food bag, and was in my tent by 9 o'clock. I texted my buddy Billy from Pittsburgh to offer condolences for the passing of his mother, and we chatted for a while. I happened to mention that I could hear an animal scratching around outside of my tent, and he responded, "Jesus! Be safe." I next called my brother David, to update him on my progress and wish him a happy 61st birthday.

Day 64
17.6 miles

Day 65 August 18, 2020 Tuesday
Stealth campsite, VA on old fire road to Blackrock Hut, VA
I woke up late due to having earplugs in, about 6:20 AM. I packed up, broke camp and was on trail by 7:15 AM. It started out as a relatively easy day, with good cool weather all morning. About a mile into the woods, I noticed one blue Croc camp shoe that someone had left lying on the ground next to a large log. I thought about picking it up in the event I ran into its owner up ahead on the trail, but just as before, I decided it best to leave it since I didn't know the direction he or she had traveled. It's a standard rule of the trail, so that if the owner returns, the item will still be where he or she left it.

I later ran into Gary from Stanimal's near the entrance to Calf Mountain Shelter. We talked about the AT and future plans for other hikes for about 10 minutes. He was planning to continue hiking north all week but considering going from north to south on each daily section. If so, we would likely run into each other every day. I also met Denitsa as she was just leaving Calf Mountain, having stayed there the previous night. We would cross paths with each other a few more times today, including where several of us would camp for the night. I would later learn that she had a daughter in Pittsburgh attending Carnegie Mellon University.

Although there were a lot of steep uphills and downhills today, I was able to make very good time on these never-ending Catoctin-age metabasalts and later the metasediments of Cambrian-age Harpers Formation. I arrived at Black Rock Hut at 5:30 and met Jeff from DC, the COO for the ACA (Obamacare). We had good discussions about Healthcare, the AT and other hiking venues. I also mentioned to him about the blue Croc shoe I found along the trail this morning. As it turned out, it belonged to him. I wished that I picked it up!

In the Shenandoahs, shelters are commonly known as huts, or less commonly, waysides. There is no difference between a shelter and a hut, but a wayside can be either a shelter, as in the case of Tom Floyd Wayside, or a roadside grill with a restroom and even gift shop and picnic area—as in the case of Elkwallow Wayside. While I was on my hike, many of the wayside grills were closed due to COVID.

Denitsa later arrived about 7 PM, very tired and exhausted from a long day carrying a heavy pack. We all made dinner together, and later she said she needed to get rid of some weight in her backpack and began offering all kinds of food to us. I agreed to take several items including a freeze-dried packet of borscht, which later I would enjoy immensely as one of my all-time favorite dinner treats. After supper, Denitsa was organizing some things in her tent when we heard her crying out something about a mouse. It turned out that a mouse had chewed through her tent, then into her backpack, going after a tiny candy wrapper in one of her backpack pockets. It amazes me how developed the olfactory senses are for mice. (I would relearn this lesson in several months while in Massachusetts).

I retired to my tent about 8:30, and shortly thereafter I heard another hiker arrive. I thought I recognized that familiar voice. It was Shutter talking with Jeff and Denitsa out at the picnic table. I finished my reading and journal entries and came out to say hello. We all sat and chatted for a while longer. Today had been very long and difficult day, and I slept well.

Day 65
17.9 miles

Day 66 August 19, 2020 Wednesday
Blackrock Hut, VA to Hightop Hut, VA

I awoke to rain about 6:30 AM. All night the rain came down and I waited until it stopped to get out of my tent and begin packing up for the day. My space blanket kept me warm but caused a lot of condensation on my sleeping bag. I had breakfast with Jeff, Denitsa and Shutter. Denitsa was hoping for a trail name, and we all made some suggestions for her. Finally, she agreed on one of Shutter's suggestions: Mama Bear. I also agreed that this was a good choice and she seemed very happy with it.

Shutter and I left about 8:30 from the campsite, and we both remarked how the trail that morning seemed easy and flat. After crossing back and forth over Skyline Drive several times, we arrived at Loft Mountain Campground and camp store around noon. We each bought several microwave cheeseburgers, and calzones for lunch. We discussed whether we should finish the day at the Pinefield Hut, a 13-mile option, or go long to the Hightop Hut, the 21-mile option. We finally decided to go for the longer mileage, and we made it to camp by 7:30. The last eight miles took us up and over several mountains, with the last climb from Powell Gap being the most challenging, perhaps because it was the last climb of the day.

For most of the day, I was transitioning from plutonic metabasalt to various Cambrian-age metasediment formations, with occasional quartzites included from the Weverton Formation. I arrived a bit before Shutter, and I set up tent in a secluded spot about 50 yards from the shelter. After I had unpacked and got organized for the evening, I returned to the picnic area. Shutter arrived about then, and we made our dinners together on the large picnic table. I decided to have the Borscht meal from Denitsa. It was one of the best dinners I had eaten on the trail since my adventures began in early June. Later I called my wife Jill for a chat, and to discuss some of the supplies I needed Steve to pick up from home. Steve is a fellow geologist and old friend who, along with two others, were planning to visit me on the trail over the weekend. It was a long day, and I was finally in my tent by 9:45 PM.

Day 66
21.4 miles

Day 67 August 20, 2020 Thursday
Hightop Hut, VA to Lewis Spring, VA

I awoke at 5:45, on High Top Mountain, to clear skies, very cool. Got up and had breakfast with Shutter, packed up and on trail by 8 AM. I got a late start, but the morning hike provided many remarkable views from the summit, leading to an easy, gradual downhill to US 33. I couldn't keep pace with Shutter and we hiked individually the remainder of the day, never to meet up again. (We would stay in touch via text messaging following our future flip-flops to Maine later in the summer, and Shutter would go on to finish his thru-hike just a few days before me). Coming down Baldface Mountain, I met Sal from Reston, Virginia, who said last night he was sleeping in his hammock and at about 2 AM, a bear came by and nuzzled his tarp but was easily shooed away.

Up ahead I noticed a clearing and soon began taking in the faint aroma of smoldering campfires. This was the Lewis Mountain campground and I ambled my way past tidy campsites and scruffy campers, on to the camp store. I walked up to the front porch, set down my pack and trekking poles, and entered the store to an organized and endless selection of packaged treats and food items. I decided I would resupply here since it was literally on the trail, and so I finished paying the attentive young cashier, left the store, and was on my way. I walked down the road to the picnic grounds carrying the bag of goodies I just purchased, and when I got to a picnic table, I took off my pack and stowed the items inside. I then continued to the trail, and after about half a mile on the flat path, I encountered a woman coming my way. We stopped to chat for a few minutes and then was on my way north again. Suddenly the trail began a steeper ascent. After taking a dozen or so labored steps, I realized I didn't have my trekking poles. My first thought was, "I must have left them back where I was talking to the last hiker I passed." But then I realized that because it was so flat and easy walking, I didn't even notice I hadn't had them with me since I left the camp store. So back I went and there they were, 10 minutes later, leaning up against the wall next to the door, just as I had left them.

Later, I was coming up onto Bearfence Mountain where a short blue-blaze trail led to the top via the Weverton Quartzite-strewn boulder scramble, allegedly affording great views in all directions. I was tired and in a rush to find a camp for the evening, so I passed on the view and kept on trucking. On my way down Bearfence Mountain, I ran into a 30-year retired Army veteran by the name of Chet who was hiking the section from Harpers Ferry to Springer Mountain. I asked him about water sources between our location and Hazeltop, where I needed to be able to find water to carry along to the summit for tonight's camp. He thought there was a nice mountain spring crossing the trail prior to starting up to the summit. I thanked him and eagerly proceeded, keeping an eye out for the spring as I headed north on the trail. I passed one meager, nearly dried-up spring on the way to the top and recall thinking to myself that this surely cannot be the spring I was looking for. It was a long but gentle climb to the summit.

When I arrived at the top, there was no water and I had not passed any other springs, but there was a nice blue-blaze path to a spacious overlook with an absolutely stunning western view. Scattered nearby were four or five handy little tent sites. This was exactly what I had been hoping for! Unfortunately for me, I had no water and could not stay there. While at the overlook, I looked down to see a stagnant puddle, complete with its own miniature floral ecosystem on one of the enormous metabasalt boulders. I briefly considered using that water for my dinner, but it was beyond grungy, and I wisely decided against the idea. I ended up hiking several miles down Hazeltop Mountain to one of the two springs about a mile before the Meadows campground. I selected a nice flat spot about 100 yards to the left of the trail on one of the old, abandoned fire roads. I set up my tent, had dinner, and turned in by 8:30 PM.

Day 67
18.7 miles

Day 68 August 21, 2020 Friday
Lewis Spring, VA to Pinnacles Picnic Ground, VA
(pick-up for motel stay)
I woke at 5:30, had breakfast and coffee, broke camp, and was on trail by 7 AM. Heading toward Meadows Campground, I met several

people who were camped out for the weekend. I chatted at length with a cyclist who had done much of the trail in the past, offering me a lot of good advice for many of the sections to the north.

While hiking, there were four deer that followed me for several miles. One buck, a doe, and two fawns kept crossing my path and running ahead of me, and then later, I would see them again. I arrived at Skyland around noon, and walked over to the Skyland Restaurant for lunch, ordered a turkey Reuben and an IPA and sat outside at a table by myself. This was so much better than my normal lunch of beef sticks, trail mix, and energy bars. I talked with two older couples, one of which I had met on the previous days hike when I left my trekking poles back at the store. We chatted for a while and then I was back on my way down the trail.

I arrived at Pinnacle picnic area about 3:30 where I was going to be meeting Steve, Robert, and Phil. I decided I would wash up in the men's restroom since I had a little bit of time before they arrived. There was a large group picnic going on at one of the pavilions, and I could smell the pleasant aroma of burgers cooking on the grill, I wanted to go over and ask for one, but I figured I could wait until dinner when the guys arrived.

At 4 PM I was sitting at one of the picnic tables with my shoes and socks off and rummaging through my pack when Steve and the others pulled up. As they exited the SUV, I could see big smiles behind their propped-up cell-phone cameras video recording the shockingly horrific spectacle of their emaciated friend, slouched over at the picnic table. We all greeted each other with wise cracks and salutations, drove back to Skyland, had drinks and appetizers, and then entered the spacious dining area for our evening feast. Steve graciously paid for my supper that evening. I ordered an enormous ribeye with all the fixings. While we were having dinner, someone stood up and cried, "Look at the bear and her cubs out there!" We all stood up and looked out the window and, sure enough, there was a mama bear and her cubs. This was the first bear that I saw since I began back in June. Later we all went back to the motel and hung out until about 11 o'clock retelling old stories, catching up on each other's lives, and talking about many of my adventures on the trail.

Day 68
14.0 miles

Day 69 August 22, 2020 Saturday
Pinnacles Picnic Ground, VA (drop off from motel stay) to Skyline Drive milepost 21.1, VA (pickup for motel stay)

I woke up in the motel at 6 AM. Today Robert was going to join me for my morning hike, which I looked forward to since I had not had many recent hiking companions. We had a quick breakfast and the guys dropped us off at the picnic grounds around 7:15. We hiked for 8.6 miles until lunch, unknowingly passing over a thrust fault at Thornton Gap, where we left the greenish-gray granodiorite of the Pedlar Formation and completed the second half of our morning stroll over younger, Catoctin-age metabasalt. Our plan was to meet up with Steve and Phil around noon at Beahms Gap. As Robert and I were coming over the last hill, we heard what sounded like two grown men trying to imitate a bear. Sure enough, as we rounded the next curve there were Steve and Phil, up to their usual shenanigans, abounding with big laughs. After lunch, I continued by myself until about 3:30 where I arrived at Hogback overlook. I called Steve to let him know where I was so he could come out and pick me up. It was a warm sunny day, and I lay down on the rock wall and had a short siesta while I waited for them to arrive.

About 4:30, Steve, Phil and Robert came by to pick me up. We went back to the motel, and after I cleaned up, we all went out to dinner at a local country restaurant. I was planning to have another big juicy steak that evening, but when I looked at the buffet bar, I decided there was a whole lot more food and went with that. We had a friendly waitress who was originally from Pennsylvania, and we talked a lot about our home state. Later, one of the head waitresses came by saying she needed to switch off the neon sign on the front window. Sitting right there, I quickly stood up and I reached into the overhead curtain box to switch off the light. The woman erupted in a raging panic and began shouting at me not to do that. By then I had already flipped the switch, and the light was off. She hastily informed me that there were live wires inside the curtain box that could have electrocuted me. "Oh," I said. "I guess I dodged that bullet!"

On the way home I said I needed to resupply, so we all went to Wal-Mart, and it took me an inordinate amount of time to find all the 10 items I needed among the half million other things I didn't need. I normally had been doing my resupplies at Dollar General, spending about 10 minutes max, but we didn't see one of those stores anywhere. The advantage here, however, was that I could purchase some Mountain House freeze-dried dinners. While more expensive than my usual ramen noodles, they are a convenient and tasty treat for a change. Phil also bought me some canned smoked trout, which he highly recommended. Later, I would concur on his astute epicurean advice. We came back to the motel, hung out until almost midnight telling stories and talking about the day over delectable convenience store wings and meatballs, ice-cold IPAs, and some of Steve's single-malt scotch.

Day 69
16.4 miles

Day 70 August 23, 2020 Sunday
Skyline Drive mp 21.1, VA (drop off from motel stay) to Tom Floyd Wayside Shelter, VA

I woke at 6 AM. We all left motel around 7 AM and went to Anthony's restaurant for breakfast. Our waitress was Mary, who was a local lass who grew up on a dairy farm near Luray, Virginia and once owned a gun shop for about 10 years. She told us how they used to make scrapple on their farm, something widely sought after by the locals. After saying our goodbyes, the guys dropped me off at the parking lot near the Hogback overlook around 9 AM.

As I was crossing the road and entering the AT, I noticed a side trail coming from farther up the road that intersected the trail I was on, and I tried to step up my pace so that I would arrive first and not be behind another group of hikers. The father and daughter coming from the other direction were moving quickly, and so I did end up hiking behind them. We marched along separately for a little while until I got close enough to say hello and meet them. Doug and Brooke were a very amicable father and daughter hiking together for a few days. We talked and trudged together for several hours, chatting most of the way. I fell behind for a little while, hiking by myself for an hour or so. During that time, I met and talked to an older gentleman about his adventures hiking the trail in Pennsylvania, and then later caught up to Doug and Brooke again, where they had stopped for lunch near a spring. We all chatted while we snacked together in the filtered sunlight of the summer forest. After finishing their lunches, the two of them said goodbye, and returned to the trail for a planned hostel stay that evening.

Later, a lady hiker stopped by and we yakked for about 30 minutes. Freethinker told me all about her thru-hike several years ago, emphasizing some of the gear she really liked, including a portable shower system from REI that she carried in her pack. She relayed stories of past years when the trail was so crowded it was difficult to find campsites by late afternoon, and often hikers would fail to dig cat holes for doing their business. That all sounded pretty disgusting to me. Following our chat, I left to continue north on my own at about 3:30. After 1.5 miles I arrived at Tom Floyd Wayside and decided to camp there for the night. It had been a short day, but I was already very tired, mostly from staying up late the previous night in Luray.

At the wayside, I talked to a young man from Princeton, New Jersey who was a math major and thinking about hiking the trail next year. I encouraged him to take some time off and just do it! Later I met a ridge runner by the name of Witt, who was out checking up on shelters and the trail in the area. Witt was a lean, athletic young man who told me that he holds the record for making it through 100-Mile Wilderness in just 34 hours. "Wow," I thought to myself. Interestingly, he accomplished this feat wearing a pair of Altra Superiors, my second pair of trail shoes carried with me every day. He told me that because they are a bit lighter weight than other Altra models, they allowed him to "feel the rocks," which he preferred. Not yet having been to that section in Maine, it was hard for me to grasp just what a tremendous accomplishment this would be. I later looked up his record, and yes, Witt "El Matador" Wisebram, surely did hold the record (1d 10h 11m 55s) for two years, until Danny "Beaver" Mejia came along in September 2020 and beat Witt's record by 53 minutes.

Later, Witt moved on and I made my dinner on the front deck of the wayside and was later in my tent by 8:30. Much, much later I too would feel ecstatic about making it through the 100-Mile Wilderness myself, however it would take me about six days.

Day 70
13.7 miles

Day 71 August 24, 2020 Monday
Tom Floyd Wayside Shelter, VA to Ashby Gap, US 50, VA

My 5:30 wake up alarm sounded, and I took longer than usual to get my stuff organized today. I made breakfast and coffee and then was on trail by 7:15. This was a big day, and I needed to get out as soon as possible. My plan was for at least 20 miles.

I stopped for my second breakfast at Jim and Molly Denton Shelter. It was almost like a private summer cabin in the Poconos, with double Adirondack chairs on deck, with a separate pavilion, and even horseshoe pits—quite extraordinary, indeed. Being encouraged by my new 100-Mile Wilderness hero, Witt, this was also the first full day that I wore my new Altra Superiors. They were super lightweight and comfortable, and I really could feel all the rocks. They had very little padding despite having a rock plate under the insole. I know they are very popular with trail runners, but maybe not the best choice for my sexagenarian metatarsals and calcanei. Coincidentally, I met three young guys from Miami and Charlotte North Carolina that day that had the same shoes as me. They all loved them.

I arrived at a shelter around 4:30 to refill water bottles. There were four or five people there but I needed to continue a little longer. One of the SOBO hikers I passed described some nice campsites ahead, and when I approached the top of the hill I saw the sign to my left, entered the side trail, and inspected several of the sites. They appeared to be nothing more than weed-whacked clearings in a nearly treeless, highbrush field, not to my liking at all. And so I continued, looking for another clearing perhaps in some patch of woods or in an open field ahead. I could find nothing that did not require a weed-whacker of my own to clear, and I was growing weary in my endless search. It was only about 15 minutes until the sun would disappear for the day, and the sky was beginning to change from blue to a pinkish orange. I finally succumbed to my exhaustion and made my camp right off the trail, next to an AT blaze post. The trail along here was a wide, grassy strip surrounded by high weeds and wildflowers in all directions, and my stealth site was merely a widening of this trail around the signpost. I would have liked to go on farther, but the sun was going down and I needed a place to set up my tent for the night. There were some large trees behind me for my bear hang, and although no water sources nearby, I had two full liters for my gloaming gastronomic gala. Later, while perched high atop the Blue Ridge in Virginia, my miraculous iPhone connected with another, located somewhere on the opposite side of our planet, and I chatted with my son Josh and daughter-in-law Vanita in Singapore for the next 15 minutes.

Day 71
21.8 miles

Chapter Thirteen
Roller Coasters, Sandy Beaches and Hang-Gliders

Day 72 August 25, 2020 Tuesday
Ashby Gap, US 50, VA to Snickers Gap, VA 7/679, VA

I woke at 5:30, quickly broke camp, had a hot breakfast, and was on trail by 7:15. Soon after my hike began this morning, I encountered a sign warning me that the "roller coaster" was about to begin. Apparently, this is a long stretch of hill climbs and descents, over a span of 20 miles or so, which holds notoriety among the AT hiking community. It seemed to also cry out to all exiting NOBOs, "See, Virginia isn't flat!"

I met several hikers along the way, mainly day-hikers or weekend backpackers. It was a hot August day, and I tried to drink plenty of water. Uphills were steep, downhills less so, but this was just in my direction. The bedrock beneath the roller coaster went back and forth between quartzite of the Weverton Formation and metabasalts of the Catoctin Formation with most ridges capped by quartzite. All these rocks were still within the Blue Ridge Complex. After climbing and descending the first seven out of the 10 mountains here, I got to the top of the seventh peak. Following the trail along the summit, I arrived at Bear Den Rocks, an expansive overlook consisting of large blocks of strongly jointed Cambrian Weverton Quartzite. It contained abundant quartz veins filling the once open joints, crisscrossing at several orientations. At the time, I thought this was the end of the roller coaster but learned the next morning that there were several more hills to complete this joy ride.

Today I was going to meet up with my new AT friend, Scott, who I had first met this summer while camping near the balds in southwestern Virginia. He and his cousin were doing a weekend backpacking trip with four of their sons and daughters, and he cordially invited me to take an evening off from the trail and visit him and his wife at their cabin in western Maryland. We met later in the afternoon at Snickers Gap, then drove back to his welcomingly rustic mountain retreat. After taking a much-needed shower and doing some laundry, I joined Scott on the front porch where we lounged and chatted while indulging in some refreshingly ice-cold local IPAs. Later, we drove to nearby Purcellville, arriving at Magnolia's at the Mill, a local dinner favorite of Scott and his wife. We were seated in the pleasantly open-air back patio, protected under an enormous awning with several other gastronomic patrons, while a blustery evening thundershower raged around us. For dinner, I enjoyed an indescribably magnificent hickory-grilled 16-ounce Gaucho Ribeye, smothered in a Chimichurri sauce, all compliments of my kind and generous friend!

Day 72
14 miles

Day 73 August 26, 2020 Wednesday
Snickers Gap, VA 7/679 to C&O Canal Towpath West Jct., MD

I woke up at 6 AM, hastily organized my gear, and came out to the kitchen where I found Scott enthusiastically preparing coffee, eggs, bacon, and English muffins for my breakfast. What a treat! I poured a cup of coffee and quickly devoured my delicious breakfast sandwich. We talked a bit more about he and his son's possible thru-hike next year, and I signed his guest log, writing a brief thank you for the hospitality and encouraging his son to hike the AT with his dad next year. In a short while we hopped in his truck and he dropped me off at Snickers Gap where I entered the trail to finish the remaining part of the famed "roller coaster."

I followed the ridgeline trail for a while, which eventually led me to a sign indicating that I had now entered West Virginia. At one of the overlooks near Raven Rocks, I met Moxy, an AT hiker originally from Spain, who did the trail back in 1996. He was quite convivial and admitted that he only realized that his name was misspelled well into his hike, and so he left it as it was. He was taking pictures with his camera and tripod and offered to photograph me standing on an overlook consisting of still more quartzite of the Weverton Formation, and then email me some of the pictures, which he kindly did later that morning. I was kind of shocked how thin and gaunt I looked!

A little while later, I arrived at David Lesser Shelter, where I sat down to have some lunch. Soon, another hiker came by to take a break, and looking fatigued, threw his oversized pack down next to mine. I said hello, and we talked for a little while. Bacon was a retired welder from Iowa who told me he was deathly afraid of bears. He also owned several Harleys and loved to ride. He seemed to have a great deal of gear packed into his oversized backpack and arrived two nights ago in Harpers Ferry where he stayed at a local hotel. He actually began his SOBO hike yesterday, but by the time he climbed out of the Shenandoah River gorge and onto the Louden Heights of the Blue Ridge, he was already out of water, exhausted, and turned back. He checked back into his hotel and called his wife to let her know he quit and would be returning home. Apparently, she convinced him to get back out there and try again. This morning was his second attempt at getting started. I didn't have the heart to tell him that he was about to enter the "roller coaster." I often wondered later how all of that worked out.

Later that day, I finally made it to Harpers Ferry by about 6 PM. I had been hiking on quartzite for most of today, but the trail was relatively easy on my feet. When I descended the trail into the Shenandoah River Gorge, I soon followed blazes to the bridge over the river. From there, it was a striking view looking east toward the confluence with the mighty Potomac. The same structural grain defining the trend of the Blue Ridge could be seen stubbornly producing lines of rapids persisting across the river. To my left, I could make out a few church steeples poking up through the trees next to the waterway. It's here that the indomitable Potomac and Shenandoah together join forces to finally erode their way through the seemingly impenetrable metamorphic barrier of the Blue Ridge. When I finally got into the town, most of the stores and restaurants were closed. I looked around to see where all the tourists were congregating, and right up the street there seemed to be a crowd. As I walked toward the gathering, I could see that there was an ice cream store still open. Excitedly I approached the front window, gazed at the menu of delightful sugary treats, and my uncontrollably carnivorous brain instantly ordered three chili cheese dogs, a large strawberry milkshake, some onion rings, and a large Gatorade. The fine young man behind the counter gave me the Gatorade at no charge, presumably because he could tell, or smell, that I was a thru-hiker.

There were lots of tourists milling around, and after finishing my meal, several people stopped to talk to me about my hike and eagerly give me useful local advice. One friendly couple told me where I could find water and recommended a campsite on a sandbar next to the Potomac River. They said to look for a nice sandy spot just after the second set of rapids, where I could set up my tent and even bathe in the river. I followed their advice, found the spot, and later waded into the shallow water to wash off the day's sweat and grit. Several passing kayakers appeared horrified at the sight of a grungy adult male bathing in their river.

The sun had just gone down, and the sky was aglow with color. I was sitting inside of my tent with my feet sticking out brushing the sand from them before closing up for the night. What I didn't realize was that in the 60 seconds or so it took me to wipe off the sand, every mosquito within half a mile radius had decided to enter my tent. While lying there on my sleeping bag reading by the light of my headlamp, I must have killed about 50 mosquitoes in the next hour. Another surprise I quickly discovered was that a train would pass by about every 30 minutes sounding its horn just in case the seismic rumble of the locomotive wasn't enough to alert me of its passing. Other than that, I slept well.

Day 73
20.1 miles

Day 74 August 27, 2020 Thursday
C&O Canal Towpath West Jct., MD to Washington Monument State Park stealth campsite, MD

I awoke at 5:30 AM on the sandy beach, which was actually a sandbar next to the Potomac River. The sand grains on my beach were eroded, transported, and deposited from all the sandstone, gneiss, and granite formations beneath the ridges cut by the mighty Potomac and its tributaries over the past millennia. These sediments are merely visiting this location on their slow but steady journey eastward toward the Atlantic Ocean and its estuaries. Millions of years from now, sandy deposits such as these may be lithified into fluvial (river-derived) sandstones, or perhaps tidal deltas near the open ocean. It was a cool, clear, and beautiful morning on the shore of the Potomac. After having a cold breakfast of oatmeal and coffee, I bode farewell to my private beach and was back on the trail by 7:15 AM.

I was looking forward to today because I would be seeing my old friend Keith from high school later this afternoon. We were both on the wrestling team for our high school, and we had not seen each other for a long time. Keith used to wear a wrestling one-piece at practice with the name "Li'l Herb," stenciled on the front, earning him that nickname by his fellow high-school wrestling comrades. The first part of my hike this morning was walking down the towpath along the old C&O canal, along the straight gravel path use mainly by bicyclists and joggers, but also part of the AT for a short section. I made several phone calls home and even talked to my son in Singapore, on FaceTime for a few minutes. I next called Keith to arrange a meet up time for later this afternoon. After a short while on the towpath, the trail abruptly turned left, went under the highway, and then began to climb up Weverton Cliffs.

The Weverton Cliffs are the outcropped expression of the Weverton quartzite, a basal Cambrian sandstone that has been metamorphosed into a very tough and resistant quartzite. Although I had been over this rock unit repeatedly in the past week, this area is the type-locality for the Weverton Formation because of its abundant outcrops in the area. I was still in the Blue Ridge, but now approaching a long ridgeline component known as South Mountain. The climb to the top of South Mountain was very steep and rocky but became less so as I neared the top. From here, trail for most of the day was relatively relaxed, with intermittent stretches of blocky Weverton quartzite littering the footpath here and there. This was obviously a popular section due to its accessibility to major roads on both ends, and I probably ran into between 10 and 20 hikers and trail runners over the course of the day. On my way out to old Route 40, there was a long straight stretch, and on my left was a large hikers' refuge, Dahlgren Backpackers Campground, equipped with tent pads, fire pits, and restrooms. Immediately after this, I came to a large parking lot next to the highway where I had planned to meet my old high-school buddy.

It was good to see Keith and we joked around for a while, made comments about each other's appearance, and laughed about the absurdity of what I was doing. I threw my pack in the back of his car, and by 3:45 we were heading for the little town of Booneville, where we soon entered Dan's restaurant on Main Street for beer and burgers. Keith explained that many of the stores, bars and restaurants in this town are owned by Nora Roberts, a renowned romance novelist who lives nearby. Dan's Restaurant was one of those. The waitress gave us a table right next to a window, far enough away from everyone that I wouldn't be offensive to any of the customers. Keith kindly, and unnecessarily, did not draw attention to the olfactory alarms going off in his brain. It was a real nice break from the trail, the burgers were outstanding, and my friend and I caught up on each other's lives, talking mostly about our families, politics, COVID craziness, and of course a bit about old times.

Later, Keith dropped me off about 6 PM at Washington Monument Park where the AT continued north. I hiked to the top of the hill and meandered through the monument and overlook area for a few minutes. The odd-shaped stone structure was one of the first monuments built in commemoration of George Washington, in 1827. I returned to the trail and continued north for about another 30 minutes, where I found a nice flat stealth campsite, and set up camp for the night. I was in my tent by 8 PM.

Day 74
20.1 miles

Day 75 August 28, 2020 Friday
Washington Monument State Park stealth campsite, MD to Falls Creek Footbridge and Campground, PA

I was up at 5:30 AM and on the trail by 7 AM. Initially the walking was smooth and easy, but soon, as I began gaining elevation, the trail became progressively rockier as the morning went on. The white blazes led me along many rugged ridgelines, which by their very nature, caused me to greatly slow my pace due to all of the debris and boulders around which I had to manage for long stretches. I had been hiking on Cambrian quartzite for about the last several days, and frequently the path would degrade to a scrambling over sharp blocks of these tough rocks.

I met several thru-hikers on my way north this morning. Some of them I would just pass with a smile and a hello, and others I would stop and talk with for several minutes. One hiker who was traveling in my direction had a slower pace than mine, and I came upon her about midmorning. We said hello and she told me her trail name was Princess Ann. She appeared a bit older than me and explained that she was originally from Cincinnati, but now living in Berkeley, California. Her goal was to make it as far north as Vermont, hopefully before the weather got too cold. She seemed quite intrigued that I was a geologist and asked me a lot of questions about rocks.

Another hiker that I crossed paths with going my way was a young man by the name of Bad Foot, a recent Penn State grad with $125,000 debt from student loans, and now working in the solar industry. Apparently, his trail name came as result of serious foot problems he suffered when he was hiking the trail last year. We talked for about an hour on our way to one of the viewpoints up ahead. He was quite verbal and intelligent, and although our political views were very different, we had a very civil and respectful discussion about a whole spectrum of topics including politics, energy, and geology.

    Later that afternoon, the trail took me to an overlook that was nothing like I had ever seen. The appearance was like something out of the imagination of Peter Max or Andy Warhol. It was known as High Rock and was at one time a popular hang-gliding spot overlooking the beautiful and pastoral Antietam Creek Valley near Hagerstown, Maryland. No longer is hang-gliding allowed here, and perhaps at least somewhat in youthful defiance, is now colorfully decorated in every conceivable type of graffiti, entirely covering nearly every square inch of exposed rock. It has apparently become a very popular local party spot for young people and aspiring closet artists. The painted surfaces afforded exactly zero opportunity for geologists to actually examine the beautiful hidden textures in these Buzzard Knob Member quartzites! I spent some time there enjoying the view, contrasting with the artwork surrounding me in every direction, and then continued north on the trail.

    The path wandered through the woods for a while before descending into a boulder field for the next two hours. I had been walking a long way today and was already tiring out. The debris field seemed to go on forever, I thought it was never going to end. This was the fourth 20-mile day out of the last five. I couldn't wait to get to camp just to lie down and rest for a while. At precisely 6:07 PM, I passed a sign indicating I had just crossed the Mason-Dixon line, meaning that I had finally entered my home state of Pennsylvania—a long awaited milestone. After another half mile, I arrived at Falls Creek Campground, where I found a terrific, flat tent-site next to Falls Creek. There were two other SOBO section-hikers there who were planning to thru-hike the AT next year. We chattered for 20 minutes or so about gear and logistics, as I set up and organized my things for the evening. A bit later Princess Ann arrived to join the three of us at our little campground by the brook.

Day 75
22.2 miles

Chapter Fourteen
North of Mason-Dixon

Day 76 August 29, 2020 Saturday
Falls Creek footbridge, PA and campground to
Rocky Mt. Shelter, PA

    I awoke in the sanctuary of my tent to the same gentle rain that had been falling all night long. It was a cool morning here in this little stream valley, but would eventually warm into a hot, muggy August afternoon. I slept in past 6:30, but by 7 AM, I was organized and ready to exit the dry warmth and transient comfort of my tent. There were three other hikers who had also camped in my same site area. Several hikers had already passed through camp by the time I got up and out for the day.

    For four of the last five days, I hiked over 20 miles, and today was a break. After yesterday's Maryland section, I was expecting more very rocky trails. To my pleasant surprise, today's trails followed a lot of low hills, and were super easy, comparatively. There was a big climb up to Chimney Rocks, a spectacular overlook that revealed breathtaking views of the valley below. While there, I met three German young men who worked in IT in New York City and Washington DC. We spoke to each other for a little while in German which was a lot of fun—with me stumbling through my limited vocabulary. I had lunch at the overlook, and shortly thereafter, the rain returned.

    I continued hiking in light rain for a while and eventually arrived at Rocky Mountain. I was again surprised at how "not rocky" it was. When I came to the turn off for Rocky Mountain at about 4 PM, I decided to at least go down and have a look. It was a steep blue-blaze downhill to the shelter, and when I arrived, there were two college girls from Chicago inside lying on their sleeping bags and reading books to one another. We said hello and talked for a short while and then I began searching for a nice, flat tent spot. The tent sites nearby were plentiful but all built on slopes, and I had to go to the bottom of the hill to find a suitable flat spot.

    While I was putting up my tent, the skies became very dark, and it looked like we were in for another big rainstorm. I got my tent up just in time as the downpour began, and the intensity of the sudden deluge was astonishing. I was quite a distance from the shelter down here, and was quickly becoming drenched, so I threw my cell phone into the tent and hastily secured the front flaps. The barrage only lasted for about 20 minutes but formed a huge lake in front of my tent. Although I was soaking wet, the heat of the day made the rain seem welcoming. I used a large stick to dig exit canals for the water in order to drain this growing body of water and prevent it from flowing toward my tent. It was actually a lot of fun and I was surprised at how quickly the water pooled.

Later the water drained, the clouds parted, and I could feel the last rays of the day's sun filtering through the wet leaves overhead. I came back up to the shelter area, made ramen noodles for dinner, talked to the Chicago gals for a while, organized my stuff, and then returned to my tent by 7:45.

Day 76
14.2 miles

Day 77 August 30, 2020 Sunday
Rocky Mt. Shelter, PA to Toms Run Shelter, PA
I was up at 5:15 AM and packed up in the dark with my lame, dim headlamp. I just had two Clif bars and coffee for breakfast, and was on trail by 7 AM, leaving behind the two college coeds eating their oatmeal. It was a steep climb out of this shelter area, and there was no sign of any more rain like I experienced yesterday afternoon when setting up my tent. It was still cool but sunny and looking like a great day ahead.

The trail was beautiful on the way to Caledonia State Park, weaving through eroded fracture passageways of massive exposures of Cambrian quartzite, some larger than my house. As I continued following the white blazes through the hardwood and hemlock forest, I passed a large group of orderly single-file young boys with their leaders, out for a couple of days of backpacking. I worked my way down to Antietam Creek to fill up my water bottles and then followed the trail up another steep but short climb over the next ridge. I caught up to Princess Ann from Berkeley again, and we hiked for a while chatting about a whole host of subjects—including geology. This topic seemed to be of great interest to her and I enjoyed our discourse. Later we had lunch together at Milesburn Cabin while sitting on benches next to a spring.

At 2 PM, I arrived at the parking lot where I was planning to meet fellow geologist and old friend Bob Heim. We worked together as geologists for a natural-gas supply company in Clarksburg, West Virginia, starting back in the early 1980s. Then in 1983, we were both moved to Pittsburgh to work for a newly formed exploration and production company, CNGD, until 1990. Bob had been waiting there for about an hour, and when I arrived, we sat by the parking lot and chatted for a while before starting out on our afternoon hike. He brought along the resupply items I had requested, including a new headlamp and food for the next few days. In addition, we feasted on some snacks, fruit, and ice-cold drinks that he brought, which we heartily consumed.

Later, it was a nice easy five-mile hike over mostly downhill terrain to Tom's Run, where we arrived by 5 PM. We passed several hikers along the way, notably two young men out for an endurance hike, wearing weighted packs and planning to hike through the night with their headlamps. When I asked why they were wearing such heavy packs, one of them responded, "So that when we finish, real life will seem easier."

Upon arrival to our camp for the night, we set up our tents right next to a mountain stream and then got dinner going on the picnic table in the common area. We met Jack, another hiker who was camped out there as well, from Oakmont, Pennsylvania. He was a retired chef instructor from a culinary school in Pittsburgh and had a very interesting cook stove. This curious little stove burned small sticks for fuel and could both cook his food and run a generator that recharged his cell phone! I had never seen anything like it before.

Day 77
19.2 miles

Day 78 August 31, 2020 Monday
Toms Run Shelter, PA to James Fry (Tagg Run) Shelter, PA

I woke up about 6 AM, had breakfast of oatmeal and coffee, broke camp, and Bob and I were on trail by 9 AM. It was a late start but that was OK as our plans were for a short day today. The weather this morning was fine, still a bit overcast, and our hike was pleasant but had odd surroundings. It appeared that all of the vegetation had been sprayed with Roundup, Agent Orange, or something, perhaps to eliminate invasive plants, (Japanese Barberry?) which were taking over. There was dead vegetation everywhere, with a sharp boundary between the healthy woods and what appeared to be a war zone.

After about another mile, we arrived at Pine Grove Furnace by 11 AM. There was a general store there known to all AT hikers for the legendary "Half Gallon Challenge." I went into the store and purchased about five days of food for my resupply, and then ordered a half-gallon of Hershey's Neapolitan Premium Ice Cream. Outside on one of the tables, I proceeded to quickly consume the whole thing in less than 10 minutes. Easy schmeasy! After finishing the ice cream, I was still hungry and decided to have a large cheeseburger for dessert. This I believe is what is commonly referred to as hiker hunger! Following the completion of this exercise in gluttony, we left for the trail, and by 3 PM, it began to rain again. We were both hiking in our shirts without rain gear, just getting soaked by the downpour. Temps were still in low 70's, and usually this did not bother me, but for some reason, today I was cold. Was it that half-gallon of ice cream I had eaten? It turns out, as I only discovered recently, that it was really only 1.5 quarts, or 0.375 gallons ice cream, not a full half gallon. I guess that's why I needed a cheeseburger to satisfy my appetite when I finished! I'm old enough to recall in the not too distant past when they actually sold half-gallon containers of ice cream. When did this change? I don't think I even realized the containers shrunk, and I'd be willing to bet the cost stayed about the same. Brilliant marketing strategy!

We arrived at Frye Shelter about 5:30 PM, and our friend Jack from this morning was already there along with another hiker. BushWacker was a retired Navy SEAL who was very friendly and never stopped talking. He was a genuinely nice guy with a big heart and he told us all about the chronic difficulties experienced by veterans due to post-traumatic stress disorder (PTSD). The four of us "old guys" got along quite well, and we had an enjoyable evening in the shelter. The rain continued all night long, and at one point an angry giant hornet, apparently attracted by the light of my headlamp, suddenly attacked me. There were a few seconds of adrenaline rush and wild panic while I tried to swat him away as he continued dive-bombing me. Somehow, I miraculously avoided being stung, and I suppose he either flew off or was mortally wounded by my insanely flailing arms. I slept well that night and stayed warm and dry in my summer liner bag covered by my trusty emergency space blanket.

Day 78
10.9 miles

Day 79 September 1, 2020 Tuesday
James Fry (Tagg Run) Shelter, PA to Yellow Breeches Creek (AT backpacker camp), PA

The four of us all woke in concert to my 5:30 AM constellation ringtone alarm dutifully sounding off in my backpack, which I forgot to turn off last night. After having been so rudely awakened too early, we all managed to get in another hour or so of morning slumber. I later arose around 7 AM. It was a slow morning due to the rain outside and with the knowledge that it was going to be a short day anyway. The four of us tentatively agreed to meet later in the day at the next shelter, Alec Kennedy, which was only about eight miles away. However, after getting out onto the trail and hiking for a while, Bob and I decided that we would probably want to continue all the way out to the Cumberland Valley, which was only a few more miles past Alec Kennedy.

The morning rain was very light, and it was a dreary overcast day. There were some great boulder crawls up Rocky Ridge and other smaller hills. We were still in mostly Cambrian-age quartzites and shales, with occasional volcanics, but still all part of the Blue Ridge province. Generally, south of the Mason-Dixon line, the westernmost ridgeline in the Blue Ridge province is known as the Blue Ridge, but that name disappears in Pennsylvania, where it becomes known as South Mountain.

It was fun hiking in the Blue Ridge through these very old rocks for the majority of the past several months. Soon however, I would emerge from Precambrian- and Cambrian-age mostly metamorphic rocks into the Cambrian- and Ordovician-age rocks of the Ridge and Valley Province. These younger rocks would be mostly limestones and shales of marine origin, and are all part of the Appalachian sedimentary basin, deposited in quiet shallow seas once covering much of the ancient continent of Laurentia. Looking down at the trail most of the time has been a great opportunity to take notice of the ever-changing rock types under my feet.

Around 4 PM, we passed by the blue-blaze trail to the shelter where the other two hikers had talked about staying for the night. We stopped briefly to reconsider joining them again, but quickly decided to continue on the AT northward to Cumberland Valley. Soon, we noticed dozens of large holes or excavations along the trail. It appeared that these were remnants of the past mining history in this area (iron ore pits) having been abandoned when the old works became too uneconomic to continue. These holes appeared very similar to the iron-ore pits near Alburtis, Pennsylvania, where my wife's Pappy Wieand used to own land adjacent to Bear Valley ski area.

By 5:30 PM, we finally emerged from the woods and into the long-awaited Cumberland Valley. We were now out of the Blue Ridge (mostly metamorphic rocks) and into the Ridge and Valley Province (sedimentary rocks). The trail led us through long stretches of soybean and potato fields, and although the rain had picked up quite a bit, it was still warm and pleasant. We continued on the trail, looking for the Appalachian Trail Hikers Campground, and finally arriving at 6 PM.

Bob and I set up our tents, unpacked gear, and then had dinner. The rain had slowed down to a very light drizzle, and I made ramen noodles with tuna, cheese crackers, and pepperoni. This probably doesn't sound very appetizing, but I was starving, and to me, this was epicurean ecstasy! Tomorrow morning the plan was for a truly gourmet breakfast at the Cafe 101 in Boiling Springs. We also soon discovered that there was an active railroad track right next to our camp, along which freight trains periodically rumbled by sounding their horns for the rest of the evening and into the morning hours.

Day 79
11.5 miles

Chapter Fifteen
Pennsylvania's Valley and Ridgeline Trails

Day 80 September 2, 2020 Wednesday
Yellow Breeches Creek (AT backpacker camp), PA to Darlington Shelter, PA

I awoke at 5:15 AM, broke camp early, and was on trail alone to go into Boiling Springs for resupply while Bob finished packing up. I bought a few days worth of goodies and food, and then went across the street for breakfast at Cafe 101. This was our third day hiking together, and my friend Bob was planning to pick up his car and go back to work after our breakfast this morning. When I got to the restaurant, I walked up to the front door, and I wondered if I would be able to go in, looking and smelling like I did. The sign at the door said hikers welcome, so I entered with my backpack, and a friendly waitress told me to come on in and I could bring my pack as well. I sat down at a corner table, waiting for Bob and looking at the menu. While I was waiting, I met Jim, another hiker who was doing a section of the AT. Soon, Bob arrived, and we ordered. I had the "Three Meat Skillet," a menagerie of eggs, home fried potatoes, and lots of ham, sausage, and bacon all mixed together in an oversized skillet plate. It was great to be eating real food! After our feast, we said our goodbyes and I began my long trek across Cumberland Valley.

The morning was overcast and foggy, the air still heavy from yesterday's rain, but the only real precipitation was the mist from the fog all around me. I was excited about today because I would be hiking across what is the equivalent of the Lehigh Valley, a Cambro-Ordovician limestone valley floor nearly identical to where I grew up just 100 miles northeast of here. Later this afternoon, I would be up on Blue Mountain, back to the rugged, familiar sandstones and orthoquartzites of the Tuscarora Formation. The trail across the valley was nice and flat, wandering through farm fields, streams, and across a lot of country roads and busy highways. I crossed over the Pennsylvania Turnpike, I-81, and Route 11, all roads that I have driven on many times before. The feeling of familiarity was stronger here than anywhere I had been on the trail to date.

I got to the Blue Mountain foothills around 3 PM. I was surprised at how easy the climb was, very well graded with some steep rock steps near the top. About halfway up the mountain, there was a rusty old car chassis with a dented front end up against an old tree. It looked like it had been picked clean of any valuable "souvenirs," and had obviously been there a very long time. The hike to the summit only took me about one hour, and by 4:30, I was at Darlington Shelter. Another hiker was already inside when I arrived, Roscoe, from New Jersey, a finish carpenter who had been on the trail for over a month. We talked for a while, and then I unpacked and set up my tent not far away. Roscoe and I began preparing dinner, and he said he was running very low on food and needed to resupply. I gave him one of my ramen packets to hold him over until tomorrow.

A little while later, Jim from the Cafe 101 arrived. We all chatted for a bit, interrogating each other about reasons for being on the trail, where we plan to go tomorrow, and things like that. Jim told us that he had been in the restaurant business operating a popular restaurant and bar for the last several years in Manayunk, Pennsylvania. He was out of work because of the devastation to his industry caused by COVID-19. I told him that he was in good company on the AT, as I had run into many unemployed chefs, bartenders, and other restaurant workers since beginning my trek in June.

Day 80
14.8 miles

Day 81 September 3, 2020 Thursday
Darlington Shelter, PA to Clarks Ferry Shelter, PA

I woke up at 5:15 AM in my tent, and everything was a bit damp from early evening heavy rains the night before. Soon, I broke camp, after having hot oatmeal and coffee for breakfast complements of Roscoe, who I had given some ramen to the night before. He was returning the favor with the coffee, which was much appreciated! I hit the trail by 7 AM, and it remained overcast all morning but was warming up as the day went on. Rain was in the forecast for this afternoon, but for now it was perfect weather to hike in as the trail led me down the mountain, into the pastoral valley between Blue and Cove Mountains.

I passed through a beautiful patchwork of fields on the rolling hills, and could see many farms, ponds, and country roads all around me. The trail continued toward Cove Mountain, and it was an easy climb to the top. I followed the trail along the Mississippian-aged Pocono Sandstone capped ridgeline, or just down slope of the ridge, all the way to Hawk Rock, a spectacular overlook. For many years, I studied these much younger rocks in the subsurface, mapping age-equivalent units such as the Weir sandstone of West Virginia, Virginia, and Kentucky from which we produced copious quantities of hydrocarbons to fuel the energy demands of our civilization. From Hawk Rock, I could see the Susquehanna River and the small town of Speeceville. After taking in the inspiring scenery for a few more minutes, I descended into the valley and soon arrived at the sleepy hamlet of Duncannon.

I followed the AT blazes through town, and my first stop was at a convenience store for some snacks, beef sticks, cheese, and Gatorade. While consuming my gastronomic treasures, I continued plodding down Market Street to "Outdoorsy," an outdoors store and hostel. I briefly considered staying for the night due to the heavy rain in the forecast, but decided to keep going, as it had not yet started. I sat down for a while on a bench in front of the outdoors store and talked with an old-time AT hiker. Scraggly, weary, and generally unkempt, he seemed really depressed and was sort of talking to himself more than to me, complaining about all his aches and pains. I made several attempts to engage in friendly conversation for a few minutes, but to no avail. I decided to move on, and after politely excusing myself, strolled a block north onto High Street. I continued for many more blocks and then followed the white blazes down a side street toward the two bridges crossing over the Juniata and Susquehanna Rivers.

Crossing the Clarks Ferry Bridge across the Susquehanna took about 15 minutes, but it was both enjoyable and interesting, amidst all the hustle and bustle of the traffic. I noticed a bright-yellow bowling ball that someone had thrown over the bridge and was clearly visible on the bottom of the river—pleading for an explanation. An odd assortment of unusual, discarded items like old shoes, car parts, and random clothing items occasionally littered my walkway. More interesting to me was the structural grain of the Upper Devonian sandstone outcrops forming the downstream pattern of rapids, at an acute angle to the flow of the river. These are some of the things you notice when you're walking and not in a hurry. After crossing both rivers, I had to cross a series of railroad tracks, and the trail then started abruptly up the mountain. When I reached one of the first terraces on my way to the top, I found myself surrounded by a grove of Pawpaw trees. I picked a ripe Pawpaw, bit into it, and found it delicious. I checked other Pawpaws that looked ripe, but all of them were still too hard and bitter to eat. These trees were common along the South Branch Potomac in West Virginia where we often canoe, but I had never seen them this far north. As I continued climbing higher, the rain returned.

My feet had been wet for quite a while, and I could tell there were a lot of hotspots and blisters starting—mostly on my toes. Continuing up Peters Mountain, I was taking in all the amazing geology such as the southeast dipping Mississippian Pocono Sandstone beds forming the linear ridgeline on this mountain. Just below the Pocono cap rock is the peculiar Spechty Kopf Formation. This rock formation contains diamictites, which are poorly sorted deposits—in this case of glacial origin. For many years, geologists speculated that these deposits were of offshore submarine fan origin known as turbidites, but more recently, most now agree that they are of glacial origin.

It was very slow going but a lot of fun scrambling over these massive rocks that I was so familiar with. The rain was now coming down very hard. The barrage followed me all the way to Clark's Ferry Shelter. When I finally arrived, there was no one else there. I waited out the rain inside for about an hour. Eventually it stopped, and the sun began peeking through the clouds. I organized my gear and decided to sleep inside under cover for the night. There were bunks on both sides, just like when the four old guys stayed together three nights ago at James Frye. I would have to wait until after dark to learn whether the mice living here were going to be active or not.

Clarks Ferry Shelter was a mere 30 miles away from my pickup location at Swatara Gap, where I would soon perform a transformation known in the AT hikers community as a "flip-flop." In a week, I would no longer be a NOBO, but instead, beginning in Maine until I finished the trail, I would morph into a SOBO.

Day 81
15.6 miles

Day 82 September 4, 2020 Friday
Clarks Ferry Shelter, PA to Stony Mt. (porcupine campsite), PA

I woke up in the shelter around 6 AM to clear skies and cool temperatures. The rain finally passed, and my feet were feeling a lot better this morning, but I wasn't sure how to deal with the emerging blisters on my toes. After mulling over my options, I finally decided to put on the brand-new Darn Tough socks and my now dry Altras, mainly for the benefit of the wide toe box on those shoes. It took longer than normal to get organized, but I was finally on the trail by 8:15.

Although the first several miles of trail were generally pleasant, I was breaking spider webs about every 20 steps—which was awful. After about four miles, I finally encountered hikers coming the other way and those spider webs vanished! I next crossed over the Route 225 footbridge, after which there was a stash of ice-cold bottled water in a cooler left by trail angels for thirsty thru-hikers. This was much appreciated since on these ridges there were virtually no water sources. From there, it remained pretty flat and easy trail most of the day.

I had an early lunch after about seven miles, when I arrived at Peters Shelter. It was a great design, with a high roof and open loft all tied into a covered front deck complete with closeable windows. There I met Poison, a SOBO thru-hiker. We talked for a while, and he mentioned there were two others behind him coming my way, Irish Bastard and Rock Steady. All three of these young men were in their early 20s, presumably just out of college.

Later as I was following the trail, it left Peters Mountain, went south and then back up on to Stony Mountain via an old forest-service road which was long, straight, and very gently graded. As I continued along the footpath, I was moving into younger-age rocks, eventually arriving near the center of a syncline where Pennsylvanian-age rocks were now beneath my feet. These rocks would contain the same metamorphosed coal beds of the once prolific anthracite region a bit further to the east, from where my maternal grandparents were born and raised. On my way up to Stony Mountain, I ran into several day- or section-hikers, including a very friendly Russian couple who were excited to hear all about my trip. They also described some cozy spots along the ridge where I could camp. When I arrived at the top, I soon found a large campground area carpeted with a thick layer of soft pine needles. To the left of the path, a hiker had already pitched his tent. I asked if he would mind my staying in the same camp area, and he said, "Sure!"

Jack was a very friendly guy from Catawissa, near Bloomsburg, Pennsylvania, and we had dinner together and talked for about an hour. Following dinner, I hung my food bag as always, and called over to ask Jack if he would like to hang his food bag on a carabineer with mine. He declined the offer saying he would take care of that himself. After dusk, when I was in my tent for a few minutes, I started seeing lights flashing around the campground, then later Jack's voice yelling at something. It turned out there were a couple of porcupines hanging around his tent and wouldn't seem to go away. This went on for quite a while and well into the night. I eventually put my earplugs in so I could get some sleep.

Day 82
16.7 miles

Day 83 September 5, 2020 Saturday
Stony Mt. (porcupine campsite), PA to Swatara Gap, PA 72 (Brother David pick-up), PA

I awoke at 5:30 AM to a very cool 50° damp morning. I slept well with my space blanket over the sleeping bag. Jack reported the porcupines never left his tent side all night. I also learned that his food bag was in his tent. We chatted over our coffee and breakfast, said our farewells, and I set off on trail by 7 AM.

The morning hike was really enjoyable, and the weather was warming up with the sun shining through the leafy canopy above. I passed several springs heading northeast toward Rauch Gap. As I followed the white blazes, the trail crossed from Stony Mountain to Sharp Mountain. I passed by many old anthracite mines along the way, as well as old settlements and mine workings with lots of anthracite coal pieces laying along trail. I met several hikers en route to Raush Gap, and we stopped to chat for a bit. Around 11 AM, I arrived at Rauch Shelter trail entrance, but it was another 0.3 miles of blue blaze to get there. I stopped for a moment to consider my schedule for today and decided just to continue northeast on the AT.

Soon after this, I passed at a sign saying, "Beaver dam ahead, expect water crossing." I curiously walked down to the end of the trail at the edge of the impoundment caused by the beaver dam. The blazed pathway abruptly disappeared under a quagmire of black, stagnant water. I was a bit surprised to see beavers active here, although they had at one time been ever present throughout most of North America until the European invasion many centuries ago. This part of the trail was already very rocky, requiring careful placement of each step, and now being underwater would make it nearly impossible for me to see where I was going. I decided against this option, and instead turned back to where the trail offered a 1.6-mile reroute along an old railroad-grade bike path, assuring the hiker of a return to the AT at its endpoint. I traveled for 1.6 miles but did not see any side trail or signage indicating a return to the AT. I continued along the detour path for several more miles, straining my eyes for any sign of a turn-off return to the AT. Eventually the trail ended at a parking lot at Gold Mine Road.

I never saw any AT return signs. Had someone removed them? Were they stolen? Had they been blown away in a storm? Was I blind? In any case, I walked south on Gold Mine Road for several miles toward Trout Run, my planned meeting spot for pick up by my brother, David. My map showed approximately five miles on road to Trout Run. Clearly, I had gotten off schedule, and needed to make up some lost time. I tried to hitch a ride with each one of the dozens of pick-up trucks that passed, but no one would stop. I thought to myself, "Welcome back to paranoid Pennsylvania." This would never happen down south!

After several miles, I arrived at Route 322, where state-park bike trails appeared on the map. It looked like from here, I might be able to get to the Swatara River, and from there, I could get to Trout Run if I followed the river trail. Eventually I found an abandoned road that was not on the map, but sort of followed the river. I came to learn that this road has been abandoned nearly 40 years ago, when the state had planned to flood Swatara Valley and create a lake. This obviously never happened. Following the old, abandoned road ultimately led me to another parking lot, but this was across Trout Run from the parking lot where David would be. I sent one last text message telling him where I was, and immediately following this, my iPhone died.

I had no idea whether or not my brother had received the text directions. I began to walk back along the road to return to our original meeting place. It was going to be about a one-hour hike, but fortunately he apparently received the message with the last nanoseconds of my battery life and was driving to my location. In a few minutes, I looked up to see his car approaching. Boy, was I glad to see that maroon SUV! Dave also brought Mom along to meet me as well as plenty of cold drinks and sandwiches. We exchanged warm greetings and then found a nice picnic table where we could all sit down and talk while I ate several gourmet ham and cheese sandwiches made by Mom. Finally, David drove us back to Allentown, arriving about 5:00 PM.

My plan was to rest and visit there for about a week, before flying to Bangor, Maine to start the third leg of my journey: a SOBO hike south from Mt. Katahdin, Maine back here to Swatara Gap, Pennsylvania. I would soon discover there would be more roadblocks ahead on my planned route!

Day 83
16 miles (estimated)

Near Thomas Knob, on way to Rhododendron Gap near Mt. Rogers, Virginia

Wild ponies grazing near trail at Grayson Highlands, Virginia

Morning fog in old fields after leaving Trimpi Shelter area, Virginia

Bear Garden Hostel, Ceres, WV

Hefty 5 foot timber rattlesnake basking in the warm sun along trail coming down Little Brushy Mountain, VA

Looking northeast into Burke's Garden from Chestnut Knob, Virginia

Breached southeastern limb of Chestnut Ridge Anticline where trail follows Tuscarora Sandstone along rim trail on Garden Mountain, Virginia

Fungi showing me his harvest of Chanterelle mushrooms at Kimberling Creek

Standing on Clinch (Tuscarora) Sandstone overlooking Wilburn and New River Valleys, near Pearisburg, Virginia

View looking northwest from Peters Mountain, near Rice Field Shelter, Virginia

Karst pastureland near Sinking Creek Valley on way to Sinking Creek Mountain

WWII hero Audie Murphy Memorial on Brush Mountain, Virginia

Thrusted remnants of Clinch (Tuscarora) Sandstone on Cove Mountain near Dragon's Tooth, Virginia

Fogged in Valley at sunrise coming down from Lost Spectacles Gap, Virginia

Overlooking Catawba Valley from Tinker Cliffs, Virginia at sunset

Standing on the iconic overhanging ledge at McAfee Knob, Virginia

On trail with Shutter in Shenandoahs near Simmons Gap, Virginia

Morning shroud of fog hanging over Shenandoahs just north of Blackrock Hut, Virginia. Periglacial deposits in foreground

Weekend trail visit with Phil, Robert, and Steve in Shenandoahs, Virginia

Start of the Roller Coaster near north end of trail in Virginia

Scott and me at his cabin near Snickers Gap, Virginia

Being treated to real food by old friend Keith in Booneville, Maryland

Graffiti-covered old hang-gliding launch pad at High Rock, Maryland

Fellow geologist and old friend Bob and me at AT midpoint, near Pine Grove Furnace, Pennsylvania

Standing on Cambrian Weverton Quartzite at Raven Rocks, Pennsylvania
Photo credit: Moxy

Navigating the knife edge of Pocono Sandstone on Peters Mountain, Pennsylvania

Part III: The Long, Cold Hike Home

Chapter Sixteen
Flip-Flop to Maine and the 100-Mile Wilderness

Day 84_A September 11, 2020 Friday
Allentown, PA (ABE) to AT Lodge, Millinocket, ME
Today was to be my travel day from Allentown, Pennsylvania to Millinocket, Maine. My brother Dave dropped me off at Allentown-Bethlehem-Easton Airport (ABE), and I flew out at 6 AM to Charlotte, North Carolina for a three-hour layover and then arrived in Bangor, Maine at 2 PM. As we were landing, I could see the welcoming landscape of these northern glacial landforms: lakes everywhere, far off mountains peaks softened by glaciers, and smoothed landscape now covered by conifers as far as the eye can see.

This was a different world, and doing a flip-flop was more than just changing directions. I was entering a new world, a new climate, full of very different flora and fauna. I was in a new physiographic province with different rocks, and there were signs of the last ice age etched into the bedrock everywhere I looked. The structural grain here was of the same northeast-southwest orientation as that of the central Appalachians, but these rocks were folded and thrusted not by the Alleghanian Orogeny, but by the much earlier Taconian and Acadian Orogenies. On top of that, the days were shorter and the skies were clearer. I had transformed from summer to autumn in one day. This was not just a continuation of my journey; this was the start of a completely new adventure!

I had made arrangements to be picked up by Paul "Ole Man" of AT Lodge in Millinocket, Maine. Ole Man and his wife NaviGator had together thru-hiked the trail 15 years ago and upon completing their trek, stayed in Maine to run their new business as proprietors of the AT Lodge. As I waited for my ride at the arrivals curb, I watched other passengers excitedly emerging from the terminal and I wondered what quests were awaiting each of them. Soon, Ole Man arrived and we began our ride back to Millinocket querying each other about our unique experiences on this marvelous trail. The tiny hamlet of Millinocket was originally built on a stream by the same name and is the last town of any appreciable size on the way toward Baxter State Park. It represents the final vestige of civilization before entering the wild and untamed north country of the Maine wilderness. Back in 2009, when my son and I stayed here, we had affectionately referred to this quaint, sleepy town as "Milli-nowhere."

When we arrived at the AT Lodge, I checked in, got new pole tips (expertly installed by Ole Man), and walked for about an hour around town. I stopped to order Mexican takeout at the AT Diner, and while I waited outside for my food, I chatted with some locals. One older gentleman told me about how Millinocket had at one time been a thriving paper-mill town, but when the factories closed, many people either left, or were just out of work, permanently. Later, I brought my food back to an enclosed common room at the lodge where I feasted alone on some of the best Mexican food I have ever eaten. I later met two hikers who had just completed the 100-Mile Wilderness and received a briefing of what was to come. We chatted for about 20 minutes, said goodbye, and I made my way upstairs to my little room for the night. Surprisingly, it was dark before 7 PM, and by 8:30, I was in bed.

Any hiker doing a flip-flop to Maine is eager to hike to the summit of Katahdin as his or her first order of business, and then move on to tackle the 100-Mile Wilderness. Of course, I was really looking forward to climbing my way to the top again, as I had done about 10 years earlier with my son. However, in the week prior, I tried unsuccessfully to make camp reservations for the day I planned to summit Katahdin. I could not reserve a tent site at either Katahdin Stream or Abol campsite. I was told by the Baxter State Park rangers to wait for two or three days or arrive at 5 AM the next morning and wait in line at the south-gate entrance to see if I could be put on a wait list. I also didn't qualify for the Birches Lean-to, due to not having come from 100-Mile Wilderness as a NOBO. Being that I was in a hurry to get started, and that my son Josh and I hiked to the summit of Katahdin back in 2009, I opted to just start at Abol Bridge, at the start of the 100-Mile Wilderness. To say I was really upset about this would be a gross understatement! So, I very regretfully did not summit Mt. Katahdin on this trip.

Day 84 September 12, 2020 Saturday
Abol Bridge, ME to Pollywog Stream, ME

I woke up at 5 AM at the AT Lodge, packed up, had breakfast in the kitchen, and left at 6:30 with Ole Man and one other rider, Michigan Millie. We were dropped off at the trailhead next to Abol Bridge about 7 AM, which officially is the start of the 100-Mile Wilderness. I never learned anything about Michigan Millie, other than that, well, she was from Michigan. We began our south trek together at Abol Bridge, and either she held back or had a much slower pace, but I never saw her again after that.

It was a beautiful day to begin a southbound journey through the lush green forests of Maine. The weather was cool, in the low 60s, and sunny with a slight breeze; very pleasant—no—perfect hiking weather.

Notably, it was now September in Maine, and most of the bugs were gone! As I rambled along this new path, I encountered a lot of NOBO hikers, mostly just finishing the 100-Mile Wilderness. For them, this was a significant milepost because Baxter was right around the corner, and soon their 2200-mile-long adventure would culminate at Katahdin Summit. All of them had that same look on their face, like a riding horse finally on his way back to the barn at the end of a long day. Later, I had my lunch at Rainbow Springs tent area, where I took off my shoes and soaked my sore, tired feet in the refreshingly cold lake water for about 10 minutes. It felt so good!

The trail was very "rooty" all day, which caused me to slow down my pace. I had to carefully place each footstep so as not to trip or stumble on the plethora of tree roots. After stopping for lunch, I encountered many more NOBO thru-hikers than that morning. Around 3:30 PM I arrived at Rainbow Stream Lean-to. Two thru-hikers were there with three other section-hikers, all just resting and snacking, but preparing to move on. Rainbow Stream was a fast-running sluice of crystal-clear water tumbling through a narrow gorge cut into the massive Devonian granite with waterfalls and deep sparkling pools all along its course.

Later at Pollywog Stream Bridge, I met Pete from Hartford, Connecticut who was also looking for a campsite. We talked to a NOBO hiker who told us about a nice campsite 1/4 mile down the AT along Pollywog Stream. I proceeded there and scouted out several false leads before finding the campsite I was looking for. It was almost 5 PM when I began the process of setting up camp. Pete arrived a few minutes later to join me, and quickly got to the business of setting up his tent. We both prepared our dinners together for the next hour, chatting all the while about our plans for the following day. After dinner, Pete extracted a small container from his pack and offered me a celebratory shot of tequila.

As the light faded for the day, four hikers wearing headlamps who planned to hike through the night approached our campsite. They all were very young and seemed excited as they explained to us about their nocturnal exercise. The sun went down here at 6:50 PM, 40 minutes earlier than where I left off the trail a week ago in Swatara Gap, Pennsylvania. I would need to readjust my schedule to wake by 4:30 or 5 AM and be on trail by 6 AM to get in a full day of hiking.

Day 84
17.2 miles

Day 85 September 13, 2020 Sunday
Pollywog Stream, ME to Potaywadjo Spring Lean-to, ME

I woke 4:45 AM. It was very cold, in the low 40s, and clear as I broke down camp and had hot oatmeal and coffee for breakfast. This was my first morning waking up after sleeping on the ground all night in Maine and it was something of a shock compared to the summer-like mornings I experienced just a week ago in Pennsylvania. When I emerged from my tent, I could see that Pete was up too, milling around his camp area getting out some breakfast items. He was only out for a couple of days and brought an abundance of food with a lot of extras that most thru-hikers would have died for. One such item was a large cache of pre-cooked bacon, which he so graciously shared with me. Before we parted, he asked me to be on the lookout today for a couple of his friends, Hank and Ollie, who did not know that he was going to be there, and let them know to stop by Wadleigh Stream Shelter. Apparently, they were either section- or thru-hiking, and he was planning a surprise appearance for them replete with an assortment of all kinds of culinary trail magic!

    I was on trail by 6:35, already recognizing this was still a late start and that I needed to begin my day earlier. About an hour into my hike, I met Music Man, a NOBO thru-hiker from Louisville, Kentucky. He enthusiastically played his stringboard instrument for me, and I captured a short video of him playing "My Old Kentucky Home." Throughout the day, I met at least a dozen NOBO thru-hikers finishing up, and on their way to summit Mt. Katahdin. All day I was questioning my decision to take the extra time to do Katahdin while I was here in Maine. I continued mulling over how I might have somehow been able to convince Baxter State Park Rangers to allow me into the park on Saturday PM to camp, or maybe have Ole Man pick me up at day's end. I had a nagging, sick feeling about not having done it when I was so close, even though I had "been there and done that" 10 years ago. In my first few days, I would be seeing Katahdin in the horizon over and over, at multiple viewpoints along the way. It seemed like every conversation I had with a NOBO included the question, "What was Katahdin like," thus prompting me to give my detailed and long-winded explanation of why I hadn't hiked to the summit on my flip-flop to become an "official" thru-hiker. Ugh!!

    Today was much harder than yesterday, and the first five miles went slow, taking me four hours to navigate the root mazes, and then climb up and over Nesuntabunt Mountain. On the way up the mountain, I ran into Hank and Ollie, and relayed the good news about their trail angel friend Pete waiting for them at a shelter up ahead. When I finally arrived at the top, there was an overlook where I could gaze across Nahmakanta Lake toward incredible views of Katahdin and other peaks to the north. This day would be a preview of what was in store for me in the weeks and months ahead. The trail was extremely rooty and somewhat rocky all day, but I still made about 16 miles, which was not bad considering my slow, late start. My goal remained set at getting on the trail by 6 AM for an earlier start.

    I arrived at Potaywadjo Spring Shelter a few hours before dusk and collected water at the springs next to the trail on the way in. Potaywadjo Spring was a wide, shallow sandy-bottomed spring with several logs and boulders arranged to allow access to the clear water. When I approached the shelter, I met Rich and Dawn, two hikers out for a long weekend with their dog, who had started a campfire in front of the lean-to. I set up my tent nearby, then came back to make dinner around the fire pit with them. Tim from Alabama was also camped down below at another tenting site and came up to join us for a bit. Rich told me about how he had attempted a thru-hike last year that ended in Monson, Maine on October 5 due to inclement weather. That date stuck in my mind as September wore on, and the weather each day became increasingly colder. He was also wearing a pair of Vibram Trek Ascent toe boots to hike and was the first person I saw on the AT donning such foot apparel. After I finished my evening feast, we all sat around the comforting fire for a while until the rain drove me back inside my tent for the night.

Day 85
15.8 miles

Day 86 September 14, 2020 Monday
Potaywadjo Spring Lean-to, ME to East Branch Lean-to, ME

I awoke 4:30 AM to my cell alarm and the sound of rain pelting my tent in erratic waves by the intermittent wind. I waited another 15 minutes before turning on the red glow of my headlamp. Slowly, over time, the rain stopped but the wind continued blowing very hard, which thankfully would help to quickly dry my tent. I packed up, had breakfast, and was on the trail by 6:40.

The morning began as an unexpectedly easy hike. I passed several hikers including a retired gentleman who had worked for 30 years at Air Products, just a stone's throw from my hometown in Pennsylvania. We talked for a while about Allentown and the trail here in Maine, and then continued on our separate ways. My plan was to stop around 10 AM for my second breakfast, perhaps at one of the many inviting fallen logs along Cooper Brook. For some reason, I decided to go a little farther to Jo May Road. This turned out to be a great decision. When I arrived there, I spotted a car parked with a table set up. More trail magic! A young lady by the name of Pineapple was making breakfast for all hikers passing through. She made me four thick slices of French toast with real maple syrup and homemade Maine blueberry topping, with a side of four sausages, and piping hot coffee with heavy whipping cream. It was all so indescribably delicious, and she also had plenty of candy, energy bars, drink mixes, and other resupply items. I stayed for about 30 minutes, along with a few other hikers, including Tinman and Bartender, who were twin brothers. They each were wearing hats bearing the letters, "TDG." When I asked what that meant they both responded, "Two-dumb guys." We laughed, and then I said my thanks to Pineapple and was on my way.

I later crossed paths with about a dozen more backpackers, roughly 2/3 of whom were thru-hiking and would be arriving at the summit of Mt. Katahdin in a few days. Around noon, I stopped for lunch on a small sandy beach along the shore of Crawford Pond. It was still a bit chilly with that same blustery wind but the scenery was magnificent. From here the trail was easy but going up and over Little Boardman Mountain was very steep, rocky, and somewhat treacherous in places. It seemed that I was still in the same granite as yesterday as I clambered over boulder after boulder on my way to the top. There were several viewpoints along the way, especially near the summit. I considered camping at Mountain View Pond, at the recommendation of Emma and Susanna who I recently met from Maine, but when I got to the pond and campsite area, the brisk wind was blowing so hard I thought it best to continue to East Branch.

When I arrived at East Branch Shelter at 5 PM, there were two women huddled together under a blanket, shivering, and another woman setting up her tent off to the side next to some large conifers. It was still very cold and windy, but this was a beautiful little camping area with several individual tent sites scattered throughout the trees, boulders, and rhododendrons. I found a quiet, secluded spot to set up camp surrounded by massive granite boulders and spruce trees, blocking most of the wind. I prepared my ramen noodles with country ham for a third night in a row—very delicious! Because of my long second breakfast at Jo May Road, I hadn't eaten much of my lunch and so I saved it for tomorrow.

At the shelter, I met Crazy Quilt who later interviewed me and another hiker, Hollywood, for her YouTube channel. It turned out that Hollywood had gone to Temple and actually knew my son's roommate, the "Hobo Documentary" guy. What a small world! Overall, today had been a very good day; I had almost made it to 20 miles. The overnight forecast was for very cold temps in the mid-30s and very windy. This would be great sleeping weather! That night, as I lay in my tent by the light of my headlamp, I dutifully reviewed my progress and planned for the next day. I was anxiously looking forward to climbing White Cap Mountain first thing tomorrow morning.

Day 86
19.5 miles

Day 87 September 15, 2020 Tuesday
East Branch Lean-to, ME to W. Branch Pleasant River (ford), ME
Awoke at 4:30 AM. After breakfast I broke camp in darkness with headlamp lit, and was on the trail by 6 AM. It was very cold this morning, low 40's and windy, but with early sun peaking through the trees. After I was on the trail for several minutes, Crazy Quilt came up from behind to say good morning. She kept pace with me as we exchanged idle chatter for a while, but she soon sped past through the boggy boardwalks and brushy

trail. I was unable to keep up with her pace and I didn't see her again until noon. Crazy Quilt was an engineer from Buffalo who, like me, had done a flip-flop earlier in the year and converted from being a NOBO to a SOBO.

The 100-Mile Wilderness can be split into two main parts. The northern half is mainly flat and rolling, weaving its way around many lakes and streams. The southern half is more mountainous and a bit more strenuous. My first big climb in part two of this wilderness wonderland was just now coming up as I approached White Cap Mountain. The rocks underlying White Cap were of the Devonian Carrabassett Formation, a series of lightly metamorphosed sandstones and mudstones. Although unbeknownst to me at the time, these rocks were age equivalent to my trail namesake, (Old) Oriskany! While on my way to the summit, I came across what appeared to be large, elliptically shaped, calcitic concretions, likely in the Hobbs or Carrabassett Formation.

The trail up to the summit was fairly steep, as I had expected, with plenty of rock steps leading ultimately to several false summits preceding the final peak—all with stunning views. There were many jaw-dropping views of Mt. Katahdin and all peaks and lakes to the north, east, and south. I ran into lots of NOBO thru-hikers. The path also led me up to West Peak, where there were several more viewpoints, but none like I had on White Cap. Later, I had lunch with Crazy Quilt and met more thru-hikers on their way to Katahdin. At one point I ran into a hiker coming from the other direction, and when I asked him about tomorrows hike over the Chairbacks, he described them as "A pain in the butt." Later, on the way down the gravel path leading to the river, I passed by the Hermitage, which is an old-growth white-pine forest covering about 35 acres next to the West Branch Pleasant River. Numerous signs announced, "No Camping Here!"

Arriving at West Branch, it was immediately clear that we had to ford the river. I sat down on a log next to the river, took off my Merrell hiking shoes and put on my trusty Altras. I secured everything else to my pack, and using my trekking poles for balance, took a few steps into the icy water. The river here was wide and shallow with lots of smooth, slippery rocks, causing me to slow my pace a bit more than I had expected. The water was frigid but felt great on my feet after a long day of hiking. Once across, Crazy Quilt and I set up our tents on nice flat areas near a small tributary to the West Branch. Later, we sat by the fire pit where we prepared our dinners and exchanged stories about the day. I ate ramen for dinner, and following that, loaned Crazy Quilt my charger so she could do some editing on her videos for her YouTube channel. Having finally crossed over the first real mountains of the 100-Mile Wilderness, I was exhausted and in my tent by 7 PM.

Day 87
16.3 miles

Day 88 September 16, 2020 Wednesday
West Branch Pleasant River (ford), ME to Long Pond Stream (ford), ME

Awoke at 4:30 AM, I packed up, made a hot breakfast, and was on trail in the dark by the light of my headlamp by 5:58 AM. It was a very slow and gradual ascent from the road, but after about an hour, the trail led me to high rock walls with white blazes going straight up the cliff face. All day was like this: reasonably well graded trails, suddenly punctuated by insanely steep climbs up walls of rock and steps. I was having a blast! I also couldn't help thinking about the guy who described the Chairbacks as "A pain in the butt." I couldn't have disagreed more! They were fun and challenging all day long, yielding great views when I arrived at each of the summits. The temperature and weather were perfect; 50s at best, but plenty of sunshine and wind.

Over the course of the day, I encountered several thru-hikers, but fewer than previous days. I met Spruce from West Virginia, an older gentleman (I learned later he was only 60, so younger than yours truly) and we stopped to chat for a few minutes. Although he was also a flip-flop SOBO like me, he moved along a bit slower, and I eventually passed him. I also ran into Crockett, the thru-hiker I met in the Smokies who scared a bear away by playing Janis Joplin on his iPod. On Fourth Mountain, I met a young man who was really interested in the geology and history of the Appalachians. He asked a lot of questions, so I gave him a quick 10-minute overview of the geology of the Appalachian Mountains. It was fun to have such an interested audience, and he seemed quite pleased and satisfied with my elucidation. Shortly afterwards, I followed the steep, craggy trail down Forth Mountain to a bog where the path now led me over a series of boardwalks. When I looked down, I noticed some strange plants clustered all around my boardwalk. At first, I thought they were Jack-in-the-Pulpits, which are common in my home state. When I bent down to look closer, I could see they were Pitcher Plants, one of the few, somewhat rare and likely endangered carnivorous plants of these unique bog ecosystems.

After climbing all of the Chairbacks, with their inspiring views and challenging ascents, I finished the morning at the summit of Barren Mountain. It was there that I found a nice rock to spread out on for lunch at about 1:30, where I could sit and gaze out to the west over lakes and expansive forested valleys. Several ground squirrels were also quite interested in my lunch, and I had to keep a careful eye on anything pulled from my pack that was edible, as I feasted on my assortment of trail food.

After lunch, I started on a long, sometimes crazily steep, shaley downhill on the Carrabassett Formation toward Long Pond Stream Lean-to. The day was pretty much continuous climbing and hiking for the last 7 1/2 hours, and this was going to be my final destination for today. When I arrived, no one else was there, and it just didn't seem the place to finish the day. I thought that perhaps there might be better campsites along the river, farther down stream. I came back to the trail and hiked down to the river where there were two young children with their mom swimming and splashing in a clear pool below a waterfall. It would later be revealed to me that they were the wife and children of Poet, owners of Shaw's in Monson, out for an afternoon swim. There, I also met Onesimus, named after a slave in the Bible who asked Paul to set him free. Onesimus was a true trail angel with a big heart, giving out candy and cookies, and he offered me an Oreo pack.

Later, I found a great campsite along the river near a small waterfall with deep pools below the falls. I set up camp, filled water bottles, and then took a fearless plunge into the icy, clear water. It was freezing cold! I hurriedly scrubbed my arms, chest, and legs, then got out and dried off. Afterwards, I hung my food bag, made ramen for dinner, and was in my tent by 7 PM. It was a great day and I was getting pretty psyched for the Whites! Overall, I would rank this day as the best so far in Maine! I seemed to keep a constant high energy level today, maybe in part because I had so much food to eat. Drinking lots of orange vitamin C mix, coffee, and energy mixes in my water probably helped too. I never did see Crazy Quilt for the rest of the day. I assumed that she ended her day at Cloud Pond, where she had planned to camp for the night.

Day 88
15.3 miles

Day 89 September 17, 2020 Thursday
Long Pond Stream (ford), ME to Bell Pond, ME

Awake at 4:30, I had breakfast, broke camp, and was on trail by 6:35 AM. I stopped 100 yards down trail to talk to Onesimus. He had more Oreos! He told me he trained as an Army Ranger and served with General Michael Flynn. He also talked about his Christianity and serving others. He really is a sincere servant and loves to help hikers with his trail magic. We talked for another few minutes, then wished each other good luck, said our goodbyes, and I was on my way.

My map indicated that there could be as many as five fords today, causing me to groan inside. As it turned out, only one river had high enough water to require fording, Big Wilson Stream. All the others allowed me to boulder hop my way across. When I arrived at Big Wilson, I changed into my Altras, but kept on my toe socks to test how they would feel when wet after crossing the river. Once I was back on dry land, I continued walking with wet feet for another two to three miles with no ill effects. At the top of Big Wilson Mountain, I changed back into dry shoes and socks. Although my feet felt fine the whole time, I have to say, they felt even better when I finally put my dry socks and shoes back on.

I tried eating more today in order to help maintain a high energy level. This meant devouring an energy bar or a bag of gorp roughly every 90 minutes. As it seemed to help me yesterday, I wanted to continue the trend. For most of the day there were lots of ups and downs, with very little elevation change. However, the trail was often very steep and seemed somewhat monotonous. It almost felt like I had gotten turned around several times and was doing the trail backwards again. I had run into a few hikers that morning, but probably the fewest of all days thus far.

I ate lunch at North Pond about 2 PM, which was one of the most beautiful places I've ever seen. Unfortunately, my cell-phone battery died, and I couldn't take any pictures. And because I also used it as a watch, I had to estimate the time. After lunch, I followed blazes for half a mile to Leeman Brook Lean-to where I helped a young lady with directions, took a short break, and then continued south. There was plenty of daylight for me to make it to the highway and even walk or hitch a ride into town to Shaw's, one of the oldest hostels on the AT. Instead, I decided to camp for the night at one of the last ponds in the 100-Mile Wilderness. This would be my last night here, and I really didn't want to leave just yet. I chose Bell Pond, where there was a nice campsite, but it was difficult to find a flat area for my tent. It took a while to set it up properly, but I finally accomplished the task.

I next made dinner consisting of ramen and tuna on the lakefront by an old canoe and watched the sun set over the pond. It was a very still and quiet evening. While I slurped my noodles, I could see no sign of animal activity, except for one or two ducks and loons. I retired to my tent by nightfall. Tomorrow I would head into Monson to rest and resupply at Shaw's. There, I could recharge my phone, resupply for the next five or six days, and perhaps buy some warm clothes for the impending trek into the Whites.

Day 89
13.5 miles

Day 90 September 18, 2020 Friday–NERO day at Shaw's
Bell Pond, ME to ME 15 (Shuttle to Shaw's), ME

I woke with first light. The morning was very cold and clear. I looked outside of my tent and could see a dense, heavy mist over Bell Pond. I quickly packed up, broke camp, had a hot breakfast, and hiked 4.4 miles to Pleasant Street blue-blaze trail, which would take me to the road leading into Monson. When I arrived at the parking lot at the head of the blue blaze trail, there was a shuttle driver dropping off a hiker. I stopped to talk to the driver, and coincidentally, it was Poet, owner of Shaw's. He offered me a ride, and we arrived at the hostel about 9 AM. He then showed me around, and I requested and then was assigned a single room. I settled in, showered, and went down the street to the general store for an early lunch.

When I returned, I met a lot of nice folks, including Potholder, Klang-on, Spruce, and Hippie Chick. I then sat outside under the shade of a large oak tree to make some phone calls and texts, and later chatted with the other hikers. At 4:30, I walked over to the BBQ place with Spruce and ordered a half-rack of ribs, mac & cheese, chili, coleslaw, and corn bread. We brought it back and ate at one of the tables inside. Excellent BBQ! Later, I talked at length with Klang-on, a NOBO who just completed the section I would be hiking in the next several days. He told me all about it, and I gave him a nice overview of what I had just seen in the 100-Mile Wilderness. I also met several other hikers, including a young man from Connecticut who was only one or two peaks away from summiting all 48 of the 4,000-foot New Hampshire peaks—barefooted! After that, I called Josh and Jessie on FaceTime, and was in bed by 10:30 PM.

Day 90
4.5 miles – NERO day

Chapter Seventeen
River Crossings, Bigelows and Saddlebacks

**Day 91 September 19, 2020 Saturday
(Shuttle from Shaw's) ME 15, ME to Moxie Bald Lean-to, ME**

It was 5:30 AM when I woke up at Shaw's. I packed up all of my gear so I was ready to go and went downstairs for a cup of coffee. When I entered the dining area, Crazy Quilt was there. She said that she had decided to pack it in for this year and restart again next year. Breakfast was at 7 AM in the dining room with the other hikers and included eggs, bacon, home fries, blueberry pancakes, and coffee. Now this was the way to start the day! The chef in the kitchen continued bringing out blueberry pancakes until everyone was full. I must have eaten eight or 10 myself! Later, one of the shuttle drivers dropped me off at the AT trailhead parking lot on Pleasant Street at 8 AM. Hoisting my pack over my shoulder, I suddenly realized I didn't have my poles. I quickly looked in the back seat and the rear cargo space and explained to the driver that I must have left them back at the hostel. I got back in the car, and we returned to find them just where I had left them, by the back porch entrance.

For the most part, the trail that morning was easy and flat, first following the edge of Lake Hebron, and then crossing the East Branch Piscataquis River. From here, it followed the West Branch for about five miles into Horseshoe Canyon. The trail going into the canyon quickly became rugged and covered in tree roots. Climbs and descents were abrupt all along the way. There were many places where the path led me steeply up to a perch overlooking the lively stream 100 feet below, then suddenly descended that same distance back to the river level. The condition of the trail remained fairly constant, in that I was navigating a spider web of tree roots pretty much the entire way.

By 1 PM, I still had not stopped for my lunch break. Suddenly, as I was approaching a gravel forest-service road, I noticed several cars parked along the shoulder, and I could smell the aroma of sausages being grilled on a charcoal fire. Up ahead, there was a small group of people standing around a smoking charcoal grill. I walked over to find that a kind older gentleman by the name of Iceman was making sausage sandwiches and offering cold drinks and other picnic treats to passing hikers. More trail magic! There were two other hikers waiting for their food, and as we talked, we soon learned that all four of us were from Pennsylvania.

My original plan for today was to climb up and over Bald Mountain after a 15-mile flat, easy section, and then camp for the night at Bald

Mountain Brook Shelter. Looking at the map and elevation profile made the day appear like a piece of cake, due to its mostly flat river trails. The problem was that the river trails required constant attention to foot placement, so that I didn't slip on a wet, smooth birch root and either fall or turn my ankle. This took much longer than expected, really wore me out, and was just a preview of what was to come.

 I made it to Moxie Bald Lean-to by 4:30, four miles short of my planned destination and decided to stop for the day. Just past the shelter, there were many nice flat campsites near the shoreline of Bald Mountain Pond. Many large, flat, car-sized boulders of granodiorite littered the water's edge. Granodiorite is similar to granite but contains less pink (potassium) feldspar and more plagioclase (calcium and sodium) feldspar than granite, giving rise to its nickname, "salt and pepper stone." Regardless, these rocks had at one time been molten and were emplaced as deeply occurring plutons back during Devonian time. More recently, for the past 2.6 million years, dozens of glacial events caused massive continental scale ice sheets to slowly form, then retreat, scouring the rock surfaces over and over again. The result of this repetitive smoothing of the bedrock surfaces was seen everywhere I looked.

 As I wandered along the shoreline, I met Vicky who had just finished collecting and filtering water. Vicky was a section-hiker from New Hampshire who had already set up her tent near an opening in the trees looking out over the beautiful pond. There was another spot next to her, where I pitched my tent, affording me the same amazing view. However, right behind me was a fallen tree, on which hung a large active hornet's nest emitting a low continuous hum. I watched as a steady stream of undisturbed insects took turns entering and leaving it. The nest didn't bother me, and of course, I didn't bother it! Two other hikers had set up their tents nearer to the shelter, JJ, a mom from the Midwest, and an older gentleman who entered his tent at 5 PM and we didn't see for the rest of the night.

 The most prevalent type of tree in this area was birch, and there were cut or fallen gray and paper birch trees lying all around. Using their abundant logs, I started a nice blaze in the fire pit, around which the three of us cooked our dinners while talking and staying warm. Birch bark contains natural oils allowing it to ignite rather easily especially when dry, and in no time, we had ourselves the start of a blazing fire. Due to my rigid, self-imposed daily schedule, I rarely took time to build fires. But having arrived at camp earlier than usual, and with the weather cooling, our little campfire made for a very relaxing and pleasant evening.

 Before dark, I needed to find a large tree from which to hang my food bag. Earlier I had seen Vicky far from our camp searching the woods while carrying her food bag, but I didn't think much about it at the time. Because of the predominance of this one tree type, I recall looking far and wide to find a good place to hang my food bag. Birch trees just don't make good bear-bag hangs. Ultimately, I had to go all the way out to the end of the blue-blaze trail, then a few hundred feet down the AT in order to find an acceptable tree. I later retired to my tent by 7:15 PM.

Stopping four miles short of today's planned destination would ultimately cost me an extra day. Had I gone over Bald Mountain to the next shelter, I might have been able to arrive at the Kennebec River to catch a ferry—about 15 miles away—by 2 PM tomorrow. The ferry only ran from 9 until 2 PM each day, and now it was unlikely I would get there in time.

Day 91
14.6 miles

Day 92 September 20, 2020 Sunday
Moxie Bald Lean-to, ME to Kennebec River, ME

I awoke 4:30 AM for an early start. It was very cold, but clear and sunny in the mid-30s. I boiled water for my oatmeal and coffee on the bank of the pond watching the eastern star-lit sky slowly transform to dawn. By 5:45, I was on the trail starting up Moxie Bald Mountain. There were several false summits, but all of them had fantastic views. At 7 AM, I reached the summit. On my way to the top, I had rustled up several spruce grouse along the trail, causing them to elicit a low-frequency wing beat that I had once mistaken for the low growl of a bear. These medium-sized game birds are similar to but more colorful than the ruffed grouse, the state bird of Pennsylvania.

My plan for today was to at least try to make the Kennebec Ferry by 2 PM. This was a huge undertaking considering I would have to hike almost 20 miles in just eight hours, which for my pace, was highly unlikely. I rushed all day long with very short breaks. After Moxie Bald Mountain, I descended into a valley, passed the next shelter, and began the ridgeline climb to Middle Mountain and Pleasant Pond Mountain. There were numerous vistas on the way to top as I climbed the bald, rocky Carrabassett backbone of the mountain. I finally reached the summit by 11:30 AM, then began my descent down to Pleasant Pond Lean-to where I had my lunch and took a short break. I really wanted to arrive early enough to catch the 2 PM ferry, but it was becoming increasingly clear it probably wasn't going to happen with 6.7 more miles to go in just 2 1/2 hours. I was exhausted and really needed a rest, so I decided to lie down in a warm, grassy area where the sun was shining. In less than 15 minutes I was fully recharged, and back on the trail.

With just a few miles to go before the river, I started looking for a campsite for the night. I didn't see anywhere at all and I arrived at the Kennebec River around 5 PM, long past my lofty goal of 2 PM. After today's marathon hike over several steep mountains and race to the Kennebec, I was exhausted. There were "No Camping" signs posted all around, but I needed to stop—and now. I set up camp on a nice flat spot about 20 yards away from the river escarpment, made Chili Mac for dinner, and was in my tent by 7 PM to begin my nightly ritual of map review, planning and documenting. Although it had been a very long, tiring day, I could sleep in tomorrow a bit later than usual since the ferry would not open until 9 AM.

Day 92
19.1 miles

Day 93 September 21, 2020 Monday
Kennebec River, ME to West Carry Pond Lean-to, ME

I woke up at 5 AM, however, in no hurry as I was at Kennebec River already. I organized my gear, packed up, and had my hot breakfast and coffee. It was still very cold, 38° overnight, and the sun was slowly coming up, which would later warm the day into the 60s. It was interesting to see that the river level had changed since yesterday.

Because there is a large dam upriver at Indian Pond, regular fluctuations in the river level are common due to releasing water out at different rates for both hydroelectric and downriver recreational considerations. During low-water season in the late summer and fall, it may appear to an approaching hiker that the Kennebec is low enough to ford. However, warnings signs are posted to caution hikers not to attempt it under any circumstances due to possibility of a sudden rise in level and the potential for drowning.

After packing up, I spent the next two hours briskly walking back and forth along the river trail to keep from freezing while waiting for the AT person to arrive at 9 AM to run the canoe ferry. At 8:45 a truck pulled up and parked near the river. I went over to talk to a convivial young man eating his breakfast sandwich in the warmth of his idling pickup-truck cab. In less than 15 minutes, he would be my canoeist to take me across the river. By 9 AM I was finally crossing the Kennebec and on my way.

Upon reaching the other side, I found myself in high grass with numerous trails running through made by the other hikers before me who had been dropped off or picked up at different locations along the bank. The tall grass and maze of trails made it hard to find the main AT here as it was poorly marked. I soon found my way and began a steady climb up the narrow, damp trail paralleling a cascading mountain stream with a multitude of magnificent waterfalls. I wanted to stop to look at all of them, but I knew I had a lot of miles to do that day, and so I kept moving. Most of the day was fairly easy grade, fast and flat, and there were several ponds—very scenic and beautiful. However, the rootiness of the trails continued to slow me down, and I kept searching for ways to increase my pace. While passing by many of the ponds and streams, I often found it much easier to just walk on the dry ground than on the boardwalks and boulders built for the wet season. At about 2:30, I came upon a very large boulder of Maine Granite on the bank of East Carry Pond where I had my lunch with the sun shining down on me. As I basked in the warming sun, it felt like summer again. Periodically a cool breeze reminded me that autumn was just around the corner. I eventually ended my brief hiatus and returned to the awaiting trail. From there, it was three more miles to the shelter that I had planned to reach that evening.

    I arrived at West Carry Pond Lean-to about 4:30. There I met Tom from Pottstown, Pennsylvania, who had already built a fire, and was busily gathering more fuel. While we were chatting, two other hikers arrived and joined us around the campfire. We all sat for awhile and shared stories of our adventures. I found a nice quiet campsite between the shelter and the privy, and just a stone's throw from the lake. Everyone else except for True North set up their tents in nearby campsites. True North was an older gentleman who recently retired from the water company in Harrisburg, Pennsylvania. He had already been on the trail for over seven months and was planning to return to his hometown of Ligonier, Pennsylvania after his thru-hike. He explained that he could not use his headlamp for the trail after dark because of the shadows cast by the rocks and roots. Just before dark, we had two more thru-hikers arrive who were on their way to Katahdin. It was a cold evening, but we all stayed warm by sitting around the crackling fire while making our dinners. That evening, I ate two servings of ramen noodles and was in my tent by 8:30 PM. I was in for a big day tomorrow and the temperature outside was very cold!

Day 93
13.1 miles

Day 94 September 22, 2020 Tuesday
West Carry Pond Lean-to, ME to Bigelow Col,
Avery Memorial Campsite, ME

    Last night, I listened to the barking and howling of coyotes for hours, and this morning, awoke at 5:30 AM for a later start than planned. Temperatures were just above freezing, and this motivated me to move swiftly. I quickly broke camp, had a fast, hot breakfast over at the shelter, and was on trail by 7 AM. The morning began with easy hiking through woods around ponds and lakes. I made it to Little Bigelow Mountain about 11:30. When I finally arrived at the exposed, windy top, I met a very attractive young lady from Maine's east coast, just enjoying the blustery day and taking selfies. She explained she was a trail runner and climbed these mountains often. We talked for about 15 minutes or so, and I thought I had almost convinced her to hike south with me a little farther, but she finally decided to go back the way she came. Standing on the bare rock face of this greenstone massif, there were fantastic views in all directions. I could see Flagstaff Lake to the north, Carrabassett Valley to the south, and the two higher Bigelow Peaks ahead to the west.

    I continued south along the trail toward Avery peak with breathtaking views all along the way. Avery Peak was named for Myron Avery who became the first chairman of the ATC in 1931, serving in that capacity until his death in 1952. Although Breton MacKaye is given credit for the original idea of the trail, and much of the earliest work, Avery was known as, "...*the person most responsible for the completion of the Appalachian Trail in the 1930s...*" according to Professor Mills Kelly and his students at the Department of History and Art History at George Mason University.

I stopped at Safford Notch Campground while just beginning the ascent, to get water. The hike to the top was long, steep, slow, and tedious, and I was tired from hiking all day. There were boulder scrambles and rock steps leading up to Avery Peak, offering occasional stunning glimpses of the Maine wilderness below. The same greenstone schist remained beneath my feet as I continued along the ridgeline to the north. It was very windy and cold, but clear and beautiful that afternoon. I finally arrived at the summit about 4 PM where there were more spectacular views.

The top was all above tree line, and very rocky with scattered tufts of rattlesnake root and dwarf red spruce—much like a Little Bigelow peak—however, much higher in elevation. From there, I could see the next summit on Bigelow, and the deep col in between which was my destination for the evening. In glacial terminology, a col is a gap or a saddle along a ridgeline, generally formed by glacial cirques carving out the mountain on either side. When a narrow knife-edge results from this same situation, the ridgeline is referred to as an arête.

I started my way down toward Avery Col Campground. It was a steep descent through a narrow boulder field of treacherous drops requiring two hands and two feet to maneuver the whole way. The wind remained strong and steady as I hiked on into the col. Upon arriving at the campground about 4:30, I found no shelter, just wooden camp platforms on one of the steep sides of the mountain. There was a boarded-up and locked caretaker's cabin near the top as well as several group tent sites. Whereas the wind was minimized on the steep sides for those wishing to camp on the platforms, I had no choice but to return to the top where I could set up and stake down my tent properly on the flat ground surface. Here the wind was steadily whipping across the mountain, probably at 25 to 35 miles per hour and it was very, very cold. It took me a while to stake and set up my trusty Altaplex using rocks to hold down the corners, so they did not blow away. Eventually my tent was up and secure with my gear inside, and I was set for the night.

I made dinner on the leeward side of the cabin shadowed from the wind so it would not blow out the flame on my cook stove. Soon after, I hung my bear bag and climbed into my tent for the night by 6:45, toasty warm inside my goose down sleeping bag.

Day 94
15.0 miles

Day 95 September 23, 2020 Wednesday
Bigelow Col, Avery Memorial Campsite, ME to Crocker Cirque, ME

Heavy winds whipped throughout the night, so I slept with my earplugs. As result, I woke up at 5:30 AM after not hearing my alarm, which went off at 4:30 AM. Because of the late start, I had no hot breakfast, just cold oatmeal and cold coffee and was on trail by 6:30 AM.

I climbed out of the col just a short vertical distance up to West Peak for yet another spectacular view of the surrounding valleys and mountains. Like yesterday at Avery Peak, everything here was above tree line, and I could see the ridgeline from Avery Peak back to the north as far as Little Bigelow Peak. It was just past sunrise, and the hazy morning sky was colored in pastels of pinks and orange. From up here, it was a long descent down to Stratton Brook and Route 27. I ran into about eight NOBO thru-hikers, and most of them were Millennials heading to Stratton for a NERO day. When I entered the woods to continue past the parking lot, I looked back to see five or six of them packing into a pickup truck. Apparently, a kind, older gentleman had offered them a ride into town.

Next was a brisk hike up to the first of several Crocker Peaks after my first lunch. On my way up to North Crocker, I sat down to have a second lunch and to carb up for next two summits. It was not a difficult climb to either peak but the dense, dwarf-red-spruce forest blocked all of my views. My hike down South Crocker had some great vistas looking south toward Sugarloaf Mountain Ski Resort. This was an extremely steep descent down the gut of Crocker Cirque. More commonly formed by alpine, or mountain glaciers, cirques are also common in these mountains as result of repeated advance and retreat of large continental ice sheets over the past 2.6 million years. The slopes of cirques are typically very steep toward the top, gradually shallowing out toward their base. My trail followed a giant crevasse filled with loose boulders of all sizes and shapes, making it a very slow descent to the bottom. My destination for the evening was Crocker Cirque Campsite, halfway down the mountain below the precipitous upper section.

Arriving at the camp area, I discovered a privy on the way in, good water, and nice flat campsites. After selecting a suitable campsite, I met another section-hiker, Nomad, from Greenville, South Carolina who was originally from Penn Hills, near Pittsburgh. We had dinner together and he gave me a spaghetti-and-meat-sauce Mountain Home dinner for which I was very thankful. More trail magic! It was a grueling day, and I was in my tent by 6:45 noticing that my inner quads were very sore from today's downhill hike.

Day 95
15.3 miles

Day 96 September 24, 2020 Thursday
Crocker Cirque, ME to Redington Stream Campsite, ME

This morning, I had a 4:30 alarm wake up, organized my gear, packed up camp, had hot oatmeal and coffee for breakfast, and was on trail by 6:09 AM. The air was cool and pleasant, and I could see the blue sky through the light fog. The start of my trek today was a gently graded descent into the Caribou Valley, leading me along the scenic South Branch Carrabassett River, and delivering me at the western base of Sugarloaf Mountain.

The sudden ascent that came next proved to be very steep and treacherous, requiring the constant use of both hands. I committed a steadfast focus on maintaining balance and secure footing on every new step of this gabbro-faced mountain. Gabbro has the same mineralogical components as basalt, but because it is cooled more slowly in the earth's interior, its crystals are visible to the naked eye. This challenging ascent was an exhilarating delight first thing in the morning. While I slowly and carefully ratcheted my body up each cliff to the next terrace, there were ever changing magnificent views over my shoulder. It was one of those mornings where I was thankful to be going up, and I wondered how NOBOs could possibly come down this type of a climb. When I reached the final terrace, it was a relatively tranquil and level ridgeline hike for the next hour or so. I opted not to take the blue-blaze side trail to Sugarloaf Summit, which would have demanded an additional half-mile and 600' vertical elevation gain to reach the radio towers and north views of the ski area.

Spaulding Mountain was a much shorter, albeit also very difficult climb, sadly offering no real views at the top. However, on the way up this craggy, sandy ridgeline, there were occasional overlooks providing views of distant ranges, as well as glimpses into Crocker Cirque. I was able to see that onerous trail to my campsite yesterday, appearing to be the main ravine running down the middle of the cirque. Following Spaulding Knob, there was more ridgeline hiking through spruce thickets. I ran into several hikers going north, and curiously, no one seemed inclined to utter as much as a hello. The path that led me down to Orbeton Stream was long and precipitous at times. Just before the bottom, the trail appeared to end at a 50-foot precipice, where I found myself looking straight down a granite cliff-face marked with occasional white blazes. Slowly I descended step-by-step, as if coming down a ladder.

I had lunch at Orbeton Stream with a young man out for a couple of days who was travelling from north to south in this area. After lunch, I climbed out of the river valley up a long steep and rocky trail to Poplar Ridge. Once on top, it was again down the other side, and then up still another backbone trek leading me to the alpine zone atop Saddleback Junior. "Funny name for a mountain," I thought. When I reached the summit, there were breathtaking views in all directions. I could see all of the peaks and ridges that I hiked in the last three days, as well as what was in store for me tomorrow. My plan for the night was to hike down the mountain and stay at Redington Stream Campsite, saving my assault on the Saddlebacks for morning. I had briefly considered continuing up and over the Saddlebacks but felt it prudent to stop for the night where I was. I had less than eight miles to go from the campsite at Redington Stream to Route 4 where I'd hitch a ride into Rangeley. The plan was to arrive at Route 4 about noon, ride into town for 1/2 day, buy fleece, Gore-Tex rain jacket and pants, resupply my food bag, and stay overnight at a cheap motel.

Day 96
16.9 miles

Day 97 September 25, 2020 Friday
Redington Stream Campsite, ME to ME 4
(Hitched ride into Rangeley), ME

My cell played its usual 4:30 AM constellation ringtone alarm to wake me up, and later I was on trail by 6:07. I relied on the light of my headlamp for first half hour of the ascent to Saddleback Horn. It was a rocky, but relatively easy climb, with several false summits on my way to the top, as is usually the case. Most of the bedrock here was granite— or more precisely—granodiorite. An abundance of muscovite mica provided a sparkling appearance to these plutonic rocks, and exposure to intense weathering resulted in a lot of grit on the trail, making for better traction for hikers.

This morning was cool but thick with fog, sadly precluding any views from the top. I was in the alpine zone, and all around me were spectacular assemblages of flora extending for miles between the horn and the main peak of Saddleback. This was a harsh environment not unlike many of the past rugged peaks I had summited since arriving in Maine. This entire mountaintop environment also reminded me very much of the Dolly Sods along the Allegheny Front in West Virginia. The bare-rock-faced mountaintops here are the epitome of the Maine tundra often described when reading about Dolly Sods.

As I slowly descended from the alpine zone back into krummholz, I could see the fog was finally lifting, but I was down into thick forest again and unable to experience those views. Krummholz is a term originating early last century to describe the severely windblown conifers present near the alpine zone. The word translated from German literally means "crooked wood." I arrived at Eddy Pond campsite around 9:11 and had my second breakfast. I sat on an old white plastic chair that had been left next to a pair of canoes on the lakeshore. Beautiful panoramas of the far-off ridges appeared as the fog lifted before me. The water was clear and chillingly cold, and I filled my water bottles before heading back to the footpath.

The hike down the mountain to Route 4 was a steady, easy downhill the whole way. The path led me to the Sandy River, and then across Route 4 to a parking lot where I arrived at about 11:15. There, I met four people who eagerly questioned me about my thru-hike, seeming rather interested in hearing about my adventures. I asked if they were going to Rangeley, and they said yes. I asked for a ride and they heartily invited me to ride in the bed of their pick-up truck into town, where they dropped me off at the Saddleback Inn. After thanking them, the driver offered me $10 saying that one of the women wanted me to have this. I declined and said I really appreciated the ride and that I should be paying you!

The Saddleback Inn was pricier than I had hoped and I was initially told it was all booked up. I later spoke to the manager, explaining that I was a thru-hiker needing a place to stay for the evening, and she thankfully came up with a room for me. I then walked down the street to a laundromat, bought food for the next few days, and shopped around for some warmer clothing. I bought a Patagonia fleece pullover at the Alpine Shop, and then had lunch at Sarge's Pub and Grub. I contacted several people about a possible shuttle the next morning but had no luck. It was looking like I might have to thumb a ride back to the trailhead tomorrow.

Later, I had dinner at the hotel restaurant, the Sunset Grill, and enjoyed a heaping portion of fish and chips. After my feast, I returned to my room, where it appeared there was some kind of an off-road ATV convention going on in the parking lot. I put in my earplugs and slept soundly.

Day 97
8 miles NERO day

Chapter Eighteen
Navigating the Notches and the Mahoosucs

Day 98 September 26, 2020 Saturday
ME 4 (shuttled from Rangeley), ME to Bemis Stream Stealth Campsite, ME

I got up at 7 AM at Saddleback Inn and got everything packed and ready to go. This was a late start for the day, and I resumed asking around for a ride to the AT trailhead on Route 4. I talked with four or five people I saw that morning with no luck and returned to my room to prepare to start walking. Suddenly there was a knock on my door. I opened it to see Paula from the front desk. She told me that a woman named Kelly offered to drive me to the trailhead for $20, but she needed to leave right away. I agreed, grabbed my pack and poles, and went outside to meet Kelly waiting by her car on the other side of the parking lot. I introduced myself and thanked her, and we left right away. We talked the whole way about lots of different topics, but mostly about her agonizing knee problems following a recent surgery.

When we arrived at the parking lot, I thanked her and hit the trail at 9:25 AM. About one mile in, I ran into NOBO thru-hiker What's Left. It turned out he was a geologist from Wisconsin who had worked earlier in his career for Hess Oil Company. We chatted for about 15 minutes or so both commenting on the great variety and diversity of rock types along the AT. He was only the second geologist I had met so far on my great adventure. Later, I stopped for a quick 10-minute break at a spacious campsite overlooking South Pond. I sat down to rest on one of the several large pine logs arranged around a fire pit, only to find gobs of pine tar sticking to me when I stood up.

My goal for today was to get to Bemus stream, past Route 17. I thought there would be water and I could camp there. I continued for a few more hours and then at about 2 PM, stopped at Sabbath Day Lake Lean-to for a lunch break. Afterwards, I filled up both liter bottles with filtered lake water and hit the trail. I followed the blazes up the side of Four Ponds Mountain where I was able to gaze out over Long Pond and several others. Today I had passed by more ponds than usual, and this really brought to my awareness the stark differences between my NOBO and SOBO journey. From Springer Mountain to Swatara Gap, none of the landscape had been subject to the continental glaciation that occurred during the last two and a half million years. The ice sheets just never made it that far south. Since starting in Maine a few weeks ago, every surface and every landform screamed of glaciation. All of the bedrock surfaces were scoured by the sandpaper effect of rock debris, or till, being dragged beneath the advancing or retreating ice sheets, creating a smoother and gentler landscape. In the south, the only ponds or lakes I encountered were dammed up streams or rivers. Some impoundments were made by man, others by our beaver friends. Here, retreating glaciers left a myriad of kettles formed by the melting of large, stranded chunks of ice, resulting in naturally occurring ponds and lakes everywhere I looked.

From Four Ponds Mountain I continued over to Spruce Mountain, and then descended into the Bemis Valley passing over Route 17 on the way. At the highway, there was a spectacular view of Mooselookmeguntic Lake. Here I temporarily lost the path after crossing the road and following what appeared to be a well-worn trail down a steep embankment toward the woods. After not seeing any white blazes for a while, I backtracked, climbing up the same steep highway embankment, only to find two people watching my futile antics and pointing me in the right direction. The man explained that he thru-hiked in the past and knew this area well. He also pointed to the white blaze right in front of me that I somehow missed before my plunge into the difficult abyss from which I had just returned. I thanked him and sheepishly slithered away, following the well-marked path.

The trail into Bemis Valley was steep and took me deep into heavily forested lowland where I crossed many dry streambeds. Finally, I saw what appeared to be Bemis Stream up ahead, but when I arrived at the wide, boulder-strewn channel, it was also completely dry. This was a sizeable stream, and it was dry as a bone. I had been hearing about how, although I was rained on nearly every day in the south, there had actually been a severe drought in New England for most of the summer. No water. Bummer! I still had about 1.2 L of water for my dinner, morning vitamins, and breakfast, which was not going to be enough.

I was very hungry when I got to camp. At about 5:45, I started setting up my tent and at the same time began boiling 1 1/2 cups of water for instant Ore-Ida potatoes. In my haste, I accidentally knocked over my cook pot, spilling most of its contents. I reacted quickly and was able to recover about 1/3 of it, but the rest had to be scraped up and thrown deep into the woods and far from my camp, so as not to attract animals. After my tent was set up and everything organized inside, I made ramen noodles with most of the remaining two cups of water, leaving only 1/2 cup for tomorrow morning's breakfast.

I didn't know what I would do for water if the springs were dry at Bemis Mountain Lean-to. The next 10 miles of trail beyond this shelter would all be on mountaintops and ridges. For now, there was nothing more I could do. I would just have to see what tomorrow would bring. By 7 PM I was in my tent, dry and thirsty!

Day 98
13.9 miles

Day 99 September 27, 2020 Sunday
Bemis Stream Stealth Campsite, ME to Sawyer Notch (ford), ME

I awoke 4:30 AM with very little water, so my breakfast consisted of a couple energy bars and cold coffee, and I was on trail by 5:55 AM. Wouldn't you know it, 100 yards up the trail I crossed a beautiful, clear mountain stream. This was Bemis Stream! I filled my water bottles while drinking copious amounts of water before continuing. I had crossed four or five dry streambeds yesterday in this wide valley before finally reaching one that was flowing. Now I was prepared for my ascent.

I started up Bemis Mountain exhilarated by the magnificent views from each of the successively higher alpine summits. This was peak New England fall-foliage season and I wondered, "Where on Earth are all of the leafers?" Perhaps they were at home, enjoying their own private showing of fall colors, virtually, on their personal computers. The white blazes painted on the bedrock led me across scoured-granite flats teaming with tundra vegetation, and scattered patches of krummholz. As I continued along the trail, each peak became progressively higher. Both the ridgeline trail and peaks scoured to bedrock with occasional tufts of highland rush, alpine bilberry, and dwarf red spruce, growing on islands of thin organically rich topsoil in between the exposed Devonian granite bedrock corridors.

I arrived at Bemis Peak by 9:30, and then scrambled my way up to Old Blue Mountain—unfortunately with no views—by 11:30 AM. Later, I dropped into Black Brook Notch and stopped for a lunch break at a small clearing next to the stream. The trail coming down from Old Blue was incredibly steep going into the notch, featuring lots of rock steps or rebar ladders cemented into the car-sized granite boulders. The trail next took me up the backside of Sawyer, then up over Moody Mountain, with the final descent into Sawyer Notch being insanely precipitous in places. The trail would lead me along contours for several hundred feet, then plunge headlong into a morass of shear rock steps and chaotically placed boulders, plummeting me quickly into a craggy abyss. What I was experiencing was a descent into a classic U-shaped valley, or notch—another unique glacial enhancement of a formerly V-shaped valley. The sides of the U-shape were the hillsides I was attempting to navigate on my way down to the stream below.

At the bottom, I found a great campsite near Sawyer Brook, complete with picnic tables, abundant flat tent sites, and a good source of water for refilling my liter bottles. Rock hopping across Sawyer proved challenging and fun, and I thankfully made it to other side without getting my feet wet.

All day I never encountered a single thru-hiker, however I did meet several day-hikers, notably a very friendly and talkative New England couple I met on top of Bemis Mountain. I recall the wife had been in the Army and was wearing a lot of camouflage. Their dog, like nearly all dogs I encountered on the AT, was also very friendly and eager to lick the salt from the skin of my offered hand.

Originally, I had planned to make it to the next lean-to but when I arrived at Sawyer Notch, the time was 4:30 and I was completely spent for the day. The additional distance was only 1.5 miles, but with a vertical gain of over 1,500 feet. I hung my damp clothing on a wash line left by a previous hiker, set up my tent, and collected and filtered two liters of water from Sawyer Brook. Dinner on my luxurious picnic table consisted of satisfyingly delectable ramen noodles with added spicey Thai tuna. Following a beautiful sunset, which painted the sky an intense shade of orange, I retired to my tent by 6:45 PM to review maps and document the day's activities.

Day 99
15.2 miles

Day 100 September 28, 2020 Monday
Sawyer Notch (ford), ME to Baldpate Lean-to, ME

It was 4:30 AM for my usual wake up, and I could feel the outside cool, but pleasant, morning air. I soon broke camp, got watered up, and was on trail by 6:15. There was a steady, light rain as I ascended up the rocky climb to Hall Mountain through thick white fog. The misty rain lasted about two hours and was soft and cool but welcoming on the uphill. Because it was morning, I still had my high energy level for a speedy ascent. I arrived at Wyman Mountain by 8:30, amongst the barren tundra landscape surrounding me. Looking off into the horizon I could see that I had been elevated above the heavy fog and misty rain, which still hung in the unknowing valleys below.

I came upon a nice trout pond on the trail, Surplus Pond, which is designated as a "State Heritage Fish Water." The pond hosts a large population of native brook trout, which are self-sustaining and are not stocked. Shortly before noon, I stopped at Dunn Falls for my lunch break, where I perched on a ledge to watch the falls cascade down into a deep notch about 100 feet below. While there, I met an older couple from Florida and we talked for a while; they mentioned how they had often thought about a thru-hike but never could make the time. Later, I continued on the trail eventually to Fry Notch, a very difficult hike downhill to the shelter area. This was one of those parts of the trail that, had I been going up, I would have felt sympathy for the poor hikers that would have to be going down—like me! When I arrived at the shelter, I noticed a sign on the outhouse reading, "Keep door closed so the porcupines won't eat the Privy."

After this I had my last big hill climb of the day, going up Baldpate Mountain. The ascent was very steep and slow going, but when I finally got to the top, there were incredible views. The trees were at their peak and leaves displayed brilliant oranges and reds as far as I could see. The trail on top was rugged and bald, marked by cairns, and open to 40 to 50 mile-per-hour winds in the near treeless col dividing Baldpate and West Peak. The trek across the col to West Peak was particularly challenging as the weight of my pack caused me to be a bit top heavy in the gale force winds. Despite the air speed, the temps were still warm, but my Indian Summer weather was about to go away very soon.

Later that afternoon, I arrived at a shelter where there was a hiker and his canine companion sitting next to a blazing campfire. Caveman and his pit bull were camped inside, and apparently had been there for some time, with no equipment and no food. Appearing homeless, I gave him some food: oatmeal, a pack of instant potatoes, a couple of energy bars, and a tuna packet plus beef jerky for his dog. They both must have been famished because all the food was totally consumed in record time. I also met a young man from Germany named Felix, who was very personable and talkative, and we spoke to each other in German for several minutes. After a long day, I was in my tent and asleep by 8 PM.

Day 100
15.4 miles

Day 101 September 29, 2020 Tuesday
Baldpate Lean-to, ME to Full Goose Shelter, ME

A 4:30 wake up, I packed up camp, made hot breakfast consisting of my standard oatmeal and coffee, and was on trail by 6:15 AM. Nice morning, foggy but warm with a light breeze. Today I would be starting my trek into the incredibly challenging Mahoosuc Range, where I would be tested like never before, and literally brought to my knees by the demanding terrain.

I had a very long hike down mountain to Grafton Notch State Park, straddling Maine Route 26. A map on the signpost showed all the hiking trails around the state park leading up to Old Speck Mountain. As I began my hike up and out of the notch and toward Old Speck, I encountered lots of rock steps going up, just as I had coming down from Baldpate this morning. The carefully placed steps made travel much easier. During my ascent, I ran into a young lady who was out for a morning hike to the summit before going to work later that day. It seemed impossible that people could do something like this before work! I can also remember making this ascent with my son, Josh back in 2009. However, I had forgotten just how long and steep the trail was to the top.

I finally arrived at the side trail intersection to Old Speck Summit about 9:30, and then continued toward Speck Pond Shelter and Campground where I had my second breakfast. The trail coming down off Old Speck Mountain was the beginning of many more along the unbelievably steep rock exposures, all too common in the Mahoosuc Range. At the shelter I refilled my water bottles and talked to a man who was spin-casting for Brook Trout at the pond. He hiked here in that morning just to fish the pond. He explained how he could tell the natives from the stocked Brook Trout because the stocked trout had clipped fins.

From there, I hiked up to Mahoosuc Arm Mountain, and then down, down, down, into Mahoosuc Notch. The descent was extremely slow going and difficult. The relentlessly steep, smoothed-biotite-granite surfaces were wet and slick from recent rains. I slipped repeatedly on the steep rock "slabs" landing hard on my butt and hips. I noticed being very sore later while hiking uphill. The abundance of birch trees firmly rooted near the slick precipitous trail proved helpful in the descent. When I finally arrived at the bottom of the notch, I crossed a beautiful, clear mountain stream. In the still of this deep, quiet gorge, I became keenly aware of the fact that there was no one else down here.

I had been looking forward to this part of the hike for several weeks, both with yearning and trepidation. The trail started out easy, but after about a mile became a rock fest. A mix of quartzite and granite boulders—some the size of your house—were everywhere. This was the mile-long trail that I had been reading about. The Notch Trail is known as the "longest mile on the AT," and it certainly proved to be for me. It took me over 2 1/2 hours to get through this valley of boulders. I had to crawl over them, around them, and under them. Navigating through the quagmire of obstructions was at times treacherous, and I traveled slowly and cautiously. Also, it was a lot of fun, but difficult and quite tedious. Toward the end, there were many places where I thought the scramble was over, only to crawl up and over another house-sized boulder to see that there was no end in sight. I was having a blast on this obstacle course, but it really wore me out!

Finally, I made the ascent to the top of South Peak of Fulling Mill Mountain, then down to Full Goose Shelter. Another hiker was already there resting inside. Fenway was from Boston, and he too was a flip-flop hiker, but arriving in Maine from McAfee Knob, Virginia. We excitedly talked for a long while about our experience just coming out of Mahoosuc Notch and what a difficult but exhilarating day we each had. Later, another hiker, Lucky Charm from Georgia, joined us. He was a thru-hiker and apparently knew Fenway from earlier in his trip. All three of us stayed in the shelter that night as it rained well into the morning. Fenway and I agreed that it would be wise to stay under cover until the rain ended the next day.

Day 101
12.0 miles

Day 102 September 30, 2020 Wednesday
Full Goose Shelter, ME to Gentian Pond Shelter, NH
I slept late in Full Goose Shelter due to heavy rain and short hiking day planned. Although yesterday was a low mileage day, it was very

demanding, but invigorating and exciting. Since we were both heading the same way, Fenway and I left camp together beginning our day about 11:45 when the rain finally ended.

Yesterday on my descent down Mahoosuc Arm, there were lots of inclined rock exposures near the "angle of repose." I like to use this term because it brings into play the effect of friction acting on an inclined surface, which either holds back my Vibram soles from slipping, or allows them to slip—in which case I go down on my backside. This critical angle differs depending on the type of rock beneath my feet and how wet the surface is. It typically varies from 20 to 35 degrees from horizontal. These rocks often contained large platy biotite-mica crystals, adding to the slick surface conditions.

My right hip was badly hurt yesterday coming down off one of the many cliffs and it was now very sore. I had four serious falls on the steep descent into Mahoosuc Notch, owing jointly to the wet conditions and the slick steep rock exposures over which I was navigating. It seemed I may have severely bruised my glutes or Iliotibial (IT) bands on both sides, and I could foresee this being with me for some time into the future. It was now painful to hoist myself up the several feet that is continuously required on these daily hill climbs, so I would have to work around the occasional discomfort, and when needed, resort to Vitamin I (ibuprofen).

There were numerous great vistas from the four mountaintops we crossed that afternoon. Although our elevations were all within 3,000 to 3,700 feet, we hiked through krummholz and alpine-zone grasses and brush for most of the day, occasionally re-entering dense, boulder-strewn spruce forests in the gaps and notches between. From Goose Mountain, we were able to see Sunday River Ski Resort to the north and east. When hiking across the bedrock-exposed highlands, we encountered many boardwalks over the ridgetop bogs, where stepping off into the murky bog could put you up to your waist in organic muck. Overall, the day was splendid with countless surprise panoramic views around every turn.

Toward the end of the day, we decided to make an attempt to reach Gentian Pond Shelter for our final destination. On the way down Mt. Carlo, we left Maine, just before beginning the climb up to Mt. Success. It was now official; we had made it to New Hampshire! We summited Mt. Success with much less effort than either of us could believe and were soon navigating our way down the last 900 feet of elevation loss, over sheer rocky drops onto multiple terraces, preceding the final approach into our shelter area. Dusk was nearing, and although we were less than a mile out, it became obvious we would soon be travelling through the woods in the dark. My headlamp lost its charge, so I had to closely follow Fenway's light the remaining way to our evening's destination. When we finally arrived about an hour after dusk, there was no one else there, and we decided to stay under cover again, rather than searching for tent sites in the blackness of the nocturnal forest. I unpacked, set up my air mattress and bag in a far corner of the loft, and prepared dinner. There was a path down to a small stream where I filled my water bottles.

When I returned, I noticed several curious mice running around the floor as we were cooking and eating, literally appearing to play hide and seek, as we prepared our evening feasts. After lights were out, surprisingly the mice did not seem to bother either of us.

Day 102
9.1 miles

Day 103 October 1, 2020 Thursday
Gentian Pond Shelter, NH to US 2, Gorham, NH

I woke up at 5 AM at Gentian Pond Shelter, sat up and noticed the truly spectacular view of the Mahoosuc Range over Gentian Lake from our trusty lean-to. My left and right hips were still very sore, and I found it difficult sleeping on either side last night. This morning's hike would follow the AT where it is shared with the Mahoosuc Trail for several miles, then up and over two mountains, Cascade and Mt. Hayes, before delivering us abruptly into the Androscoggin Valley near Gorham, New Hampshire. After Gorham was the start of the White Mountains, for which I needed to be in prime condition. I was happy that today's was going to be a comparatively short hike to Rattle River Hostel, where I could give my feet, knees, and hips a much-needed healing and rest for a day or so before tackling the Whites.

Fenway and I started out heading down to Gentian Pond, and then followed blazes up onto a terrace that led us to the upper pond. When we arrived at Dream Lake, my right hip was really bothering me. After mulling over the map for a few minutes, we decided to take Peabody Trail to the hostel instead of following the AT over the last two peaks. It was not a significant detour, but Peabody Trail followed a nice stream with lots of beautiful waterfalls along its course and was also much smoother and more gently graded.

We arrived at Rattle River at 11 AM. After checking in and meeting some of the folks there, we both decided to take a ZERO day tomorrow and then slack pack 21 miles from Pinkham Notch back to Rattle River Hostel the following day. After hanging my pack in the mudroom, I took a steaming hot shower and did some laundry. At noon, Tigger shuttled Fenway and me into town for some lunch. We stopped at a barbecue place for some mouthwatering brisket and local IPAs, and later we came back, relaxed, and watched some movies. I talked to some other hikers, and later met "Jim the Golf Guy," who told me all about his work with the local golf industry. Jim was genuinely a nice fellow, a natural salesman, and truly had the gift of gab. It was getting late, and

I had a long day, and needed to get to bed. My feet hurt and my leg and hips were stiff and sore. I was already worrying about the 21 miles on Saturday. I really needed to rest!

Day 103
11.8 miles

Day 104 October 2, 2020 Friday
US 2, Gorham, NH to US 2, Gorham, NH
    ZERO day. I slept in late, up at 8:30 AM and later came down to the kitchen and made a nice batch of pancakes for breakfast. Later, Fenway and I got a shuttle from Tigger and Eric into North Conway to re-supply and stop at REI for Gore-Tex, new trekking poles, fuel canisters, and several other items. There was a line in front of the REI where only a few shoppers were being allowed into the store at one time. On the way back, we stopped for lunch and had a sandwich and a cold IPA. The remainder of day was just hanging out and resting at Rattle River. Later, we arranged for tomorrow's 21-mile slack pack of the Wildcats. It would be so nice hiking these mountains—if for only one day—with a light pack.

Day 104
0 miles–ZERO day

Chapter Nineteen
The Whites

Day 105 October 3, 2020 Saturday
US 2, Gorham, NH to Pinkham Notch Visitors Center by shuttle, NH

    6 AM wake up in Rattle River Hostel. I went downstairs for quick breakfast and coffee. We took a shuttle to Pinkham Notch, arriving at 7:00, and were on trail by 7:15. Fenway and I were slack packing today, but from south to north so we would then be able to walk back to the hostel without the aid of another shuttle. On our planned hike, we would summit ten 4,000-foot peaks over 21 miles, and hopefully get back to the hostel before dark.

    The geologic history of White Mountains, although complex, is mostly about two collisional events, the Taconian Orogeny about 450 million years ago (Ma), and more importantly, the Acadian orogeny, about 360 Ma; followed by two major times of much younger magmatism, first beginning about 200 Ma, and the second around 130 Ma. The result is a collection of many types of highly deformed metamorphic rocks and igneous rocks that mostly cooled and solidified beneath the surface then slowly exhumed over geologic time. The rocks on which we would be hiking in the Wildcats, Carters, and Moriahs were primarily gneisses of the early Devonian Littleton Formation.

    We began our trek by following several streams for the first mile, which was fast and easy. The blazes then quickly led us straight up countless cliff faces and along a rugged ridgeline for the next 1.5 miles to Wildcat "D" Peak: an elevation gain of 3,000 feet. At the summit, I passed by the top of the quad chairlift for Wildcat Ski Resort. We passed many groups of day-hikers out for a weekend in the White Mountains. Some of the locals we talked to commented that they considered Wildcat to be one of the best ski resorts in New England.

    By noon we were already at Carter notch, an 1,100-foot-deep, very steep, and rocky descent into the glacial gash between Wildcat Mountain and Carter Dome. We stopped for our lunch break at Carter Notch Hut, which was boarded up and closed for the season. Climbing out of the notch was equally steep and treacherous, but I made short order of that passing just about everyone and making it to the top in about 30 minutes. That rest day yesterday seemed to be really paying off—the average day-hiker's pace could not keep up with mine.

    There were endless incredible views over much of the day along the many open-ridgeline trails connecting all ten of the 4,000-foot peaks. The morning brought mostly clear skies with occasional cloud cover. Later, some heavy clouds moved in obscuring views and actually bringing snow to the top of one of the Wildcat peaks. It was an unusually festive mood and many of us found ourselves singing Christmas carols as we rambled along in the falling snow. I was continually running into day-hikers and talking to lots of people throughout the day. Fenway and I traveled at different paces at times, often hiking alone for an hour or two, and then one of us would eventually catch up to the other.

By 6:45 PM it was almost dusk, and we still had 4.5 miles to go. Fortunately, we did bring our headlamps to be able to see the trail in the dark. We were also fortunate that those last miles were on very good trail—all downhill and nicely graded most of the way. We finally made it to the asphalt road, and eventually arrived back at Rattle River by 8:15 PM. I ordered spaghetti and an antipasto salad from one of the local Italian restaurants for dinner and it was excellent.

Day 105
21.1 miles slack packed

Day 106 October 4, 2020 Sunday
Pinkham Notch Visitors Center, NH to Osgood Tent Site, NH

I woke up late this morning after yesterday's marathon 21-mile slack pack through the Wildcats. Today was the final morning at Rattle River, and it was time for Fenway and me to head toward the Presidential Range—once again with full packs. Our plan was to just hike into a campground area near the base of Mt. Madison, and then begin early the next morning and get as far as we could through the Presidentials before dark.

We packed up and left Rattle River by 10 AM, and a shuttle dropped us off at Pinkham Notch Visitors Center where we had lunch before setting off on our trek. Tourists, hikers, and photographers were everywhere on this busy Sunday, since Pinkham is the location for many of the trailheads into the Whites. We were on trail by 11:30 AM, with only 4.8 miles to Osgood tent sites. It was a fairly easy hike and by 2:30 we arrived at Osgood. Fenway's knee was bothering him, so he was finished for the day. I briefly considered continuing on up Osgood Trail to Mt. Madison, and then looking for a campsite off the mountain somewhere. Yesterday I had talked to a local hiker while at Rattle River who told me about some of the options I would have for camping, were I to do this. After giving each alternative some thought and weighing the possibilities, I decided that it was best to just stay here for the night and hit the Osgood Trail with fresh legs first thing in the morning.

I worried a bit about the travel distance for tomorrow. I was planning to shoot for 15 miles, which was perhaps more than I could do up there. I set up my tent, made dinner on one of the tent platforms, and then turned in for the night. The temperatures were dropping already into the low 40s with an expected low in the high 30s. I would plan to get off tomorrow morning by 6 AM in the dark with my headlamp.

Day 106
4.8 miles

Day 107 October 5, 2020 Monday
Osgood Tent Site, NH to Mizpah Spring Hut, NH

After a 4:30 alarm wake up, I packed up my gear, had a hot breakfast and coffee, and was on trail by 5:55 AM with my headlamp blazing the way. Days were becoming noticeably shorter by about three minutes each day. Being thrust into much higher latitudes since arriving in Maine, this had been both intriguing and frustrating at the same time. Whereas daylight loss was about 2 1/2 minutes per day in Pennsylvania, it suddenly became over three minutes per day since arriving in Maine less than one month ago. Not much, I thought at first, but cumulatively, I was losing one half hour of daylight every 10 days, or roughly about an hour every three weeks. This effect would be greatly amplified in early November when we turned the clocks back to EST. Suddenly, if I wanted to get in those same number of daylight hours before dark, I would need to begin my day at 3:30 AM. Ugh!

I checked with Fenway on my way out and he said his knee was still hurting so he would try to catch up later in the day. It was cold but pleasant, 39° and slightly cloudy. Osgood Trail up to Mt. Madison turned out to be much easier than expected and I was pleasantly surprised. It took me approximately 2 1/2 hours to climb the 3,000-foot vertical to the top. There were remarkable views once I entered the alpine zone at about 4,400 feet, with lots of patches of sunshine, but dense fog hanging in all of the valleys. I looked back occasionally but saw no sign of Fenway. I met several people on Mt. Madison, at the Madison Hut, and on later peaks or trails throughout the day. Several hikers said they passed him, but it seemed he was still pretty far behind.

Throughout the entire day, there were never-ending panoramic views with clear to partly cloudy skies and excellent visibility. I was able to see Wildcat Ski Resort to the southeast, and Breton Woods Ski Resort off to the southwest. Mt. Madison Summit was at 5,366', but the highest elevation in the Whites was about six miles ahead on Mt. Washington at 6,288'. Once above tree line, following the white blazes pretty much just meant rock hopping over more Devonian-age metamorphic gneiss and schist boulders, all the way to the summit of Mt. Washington. It amazed me just how far I could see from up here with no tree cover whatsoever along this ridgeline. My view of the trail ahead extended all the way out to beyond Mt. Pierce, almost 12 miles from where I was standing.

From Mt. Madison, the AT followed the Gulfside Trail all the way to Mt. Washington. This mostly ridgeline trail wound down to the Madison Hut, located in the col between two peaks. Then, rather than taking hikers up and over Mt. Adams, it bypassed the peak, hugging the west side of the mountain, then returned to the ridgeline after Adams. Hiking became less difficult after Mt. Washington. Large, jagged boulders were a common feature along the Gulfside Trail, so it was essentially all boulders up to Mt. Washington summit. Thereafter, the trail abruptly transformed to steps or gravel path with occasional rock scrambles. This more refined section south of Mt. Washington, known as Crawford Path, was built by Abel Crawford in 1819, and according to the USFS, is "The oldest continuously used mountain trail in America." The AT followed the Crawford Trail from Mt. Washington to Mt. Pierce (still called Mt. Clinton by some) for about 5.3 miles. Initially I thought the name change was honoring our 42nd president, but it turns out the first name given to the mountain was in honor of a former New York Governor, DeWitt Clinton. In 1913, it was renamed after Franklin Pierce.

As I gradually approached Mt. Washington, its much greater stature and prominence became increasingly evident. The summit towered over anything else visible in any direction. It was massive. I had been to the summit in 2009 with my son Josh, when we attempted a one-day 17-mile loop starting at Pinkham Notch, climbing up Tuckerman's Ravine, then following the Gulfside trail to Madison Hut, and finally back to Pinkham via the Great Gulf Wilderness. It was a valiant attempt, but we never made it quite all the way before darkness set in. I can still recall hiking well after 9 PM in the Gulf, in the dark with our headlamps on, looking for one of those little campsite markers on a tree. We finally did find a campsite, set up our tents, and even made a small campfire to cook dinner before retiring to our tents by midnight. All this to say, at 2 PM when the Gulf Trail brought me within 0.2 mile of the Mt. Washington summit, and the side trail pointed up to the peak, I just followed the white blazes, staying on the Gulf Trail and thinking to myself, "Been there, done that!"

There were literally car-loads, and train-car-loads of tourists on and around Mt. Washington. I watched several times as the cog-railway cars chugged up the steep track leading to the summit. Built in 1869, this unique mode of transport to the summit boasts as the second steepest railway in the world. Its once coal-powered steam engines are now advertised to ecofriendly tourists as "running on biodiesel." I could still see chunks of coal from bygone days, strewn all along the path of the railway as I passed by. The other large group of tourists on the summit arrived by car, truck, or van, via the Auto Road. Most years a person could hear languages from all over the world spoken here, but aside from a few Asian tourists, most of the people looked and sounded like locals. After Mt. Washington, on my way down to Lake of the Clouds, I met "Abe Lincoln," a two-time thru-hiker, currently on his second SOBO, which began September 19. For the next several minutes, we walked together while exchanging information about ourselves. He was from Tennessee and worked for a time as a river-rafting guide in West Virginia on the New and Gauley Rivers. He preferred the fall and early winter for his thru-hikes because there were fewer people, snakes, and bugs on the trail. Abe was long and lean, his pace much faster than mine, and he soon left me in the dust.

It had been a very long and strenuous day, but now that I had passed Mt. Washington, everything was generally downhill from here. The southbound AT in this area continued to sidestep many of the peaks to the east or west. The Crawford Path and AT bypassed both Mt. Monroe and Mt. Eisenhower, and it was not until Mt. Pierce that the trail actually followed a ridgeline to the summit. The path to the top was very steep and followed wide swaths of exposed gray gneiss of the Littleton Formation, where shimmering flecks of both muscovite and biotite mica could be seen all around me. My target destination for that evening was Mizpah Hut past Mt. Pierce, where there were designated campsites at the Nauman tent-site area. The last quarter mile of the descent to Mizpah was brutal. It appeared a microburst or tornado leveled a 50-foot-wide swath, which I followed for about a quarter mile. Within the swath, nearly all of the trees were blown down along this extremely abrupt and treacherous grade. I arrived at the hut and campsites around 5:30 PM, set up my tent, unpacked, got water, and made beef stew for dinner. When I walked over to stow my food bag in the bear box, I noticed several hikers sitting around a fire. I said hello, and briefly considered joining in, but was just too tired. I saved a nearby flat tent-spot for Fenway, but he never showed up. I was in my tent by 7 PM. I had done less than 15 miles today, but here in the Whites, carrying a 40 lb. pack was a hard day's work. I felt relieved to be over the hump!

Day 107
14.8 miles

Day 108 October 6, 2020 Tuesday
Mizpah Spring Hut, NH to Ethan Pond Shelter, NH

I awoke at 5 AM, stiff and sore from yesterday. After packing up, I had a hot breakfast and coffee and was on trail by 7:30 AM. It was a late start, but that was intentional. Weather this morning was cool but partly sunny. It looked like a great day ahead, but I put on my woollies and Gore-Tex pants in the event of rain, as the forecast was warning of rain at 2 PM. Today's hike was over several more peaks and then down into Crawford Notch, then back up to higher elevations again afterwards. It was difficult terrain over Mt. Jackson and Webster, and very slow going. My hip was still hurting from that third fall in the Mahoosucs. Also, my right knee was acting up on the downhills. Overall, I just felt beat up from 15 miles on the Presidentials yesterday. I even considered getting to the highway and going in for another ZERO or NERO day.

The next six miles out to the highway took me six hours. It was a remarkably slow day. The area around Webster Cliffs was scenic and spectacular, with continuous cliff walks for well over an hour. Were my legs in great shape, this would be fantastic! I changed into shorts just before Crawford Notch and felt noticeably cooler and much better. I also ate lunch. Maybe feeding the hunger beast was the primary reason for feeling better, I don't know. I crossed over Highway 302 at Crawford Notch and walked up the road to a parking lot for the AT trailhead. There I met lots of friendly tourists and day-hikers and had several nice chats about my thru-hike. Perusing the license plates of all of the cars indicated tourists from all over had come to hike through these colorful mountains today. Behind the parking area I could see railroad tracks. A young lady was waiting there with her camera ready, as a train approached. I took a short video of the train, all passenger cars with lots of waves from open windows, and lots of COVID masks. Each train car had a woman's name written on the side, including one car bearing the name, Dorthea Mae. Ironically, *Dorothy* Mae was the name of my maternal grandmother.

The trail to Ethan Pond was fairly easy and it was just three miles up to the shelter area. I arrived to a very "Maine-like" setting, hosting a beautiful alpine pond with glacially sculpted mountains in the background. There was no sign of rain, but lots of puffy clouds with sunrays shining through. During that afternoon, the weather became noticeably cooler and the winds began picking up. As I was retrieving my stove and food from my pack to start on dinner, I noticed someone coming up the blue-blaze approach path. It was Fenway. It turned out that he was able to make it past Mt. Washington yesterday, but only as far as Lake of the Clouds Shelter. He then hiked 15 miles today to reach this location. We talked and planned for the next couple of days of hiking together. We still had the Franconia Ridge, Kinsman Range, and Mt. Moosilauke ahead of us. I projected about one full week before crossing from New Hampshire into Vermont. Somewhere near Hanover, New Hampshire seemed like a good next hostel town for a full day of rest.

Day 108
9.3 miles

Day 109 October 7, 2020 Wednesday
Ethan Pond Shelter, NH to Garfield Ridge Shelter, NH

I was up at 4:30 AM, went over to bear box to retrieve the food bag, and made quick breakfast. By 5:55 am, I was packed and ready to hit the trail. Fenway decided to hang back for a later departure. It was still dark so I wore my headlamp for about the first 30 minutes. Ethan Lake was beautiful before the sun came up, and I stopped to take in the essence of the predawn for a short while before I continued down the path.

It was an easy trail that morning all the way to Zealand Falls Hut. When I arrived, two AT workers were winterizing the hut, Footloose and Lucy. They said the doors would open later that day for coffee and snacks but for now it was work time. About 10 yards away, there was a beautiful mountain stream cascading over the Jurassic granite bedrock in a spectacular series of waterfalls. These are some of the youngest rocks found on the AT and resulted from the breakup of Pangaea 200 million years ago. There was a working water spigot next to the stairway, so I filled my liter bottles and talked for a while with Footloose. To my pleasant surprise, he told me that the trail up to South Twin Peak was pretty flat and well groomed. It turned out he was absolutely correct.

The ascent to Zealand Ridge was steep but nicely graded, and soon I made it to the top. I followed the ridge trail to Mt. Guyot where there were some magnificent viewpoints. At one of the overlook ledges, there was a large bundle of old railroad ties that apparently had been air-dropped there by helicopter. Trail workers would likely use them to build new steps or waterbars along the trail. As I continued on the ridgeline trail toward South Twin Peaks, the skies grew dark, and it began looking like rain. After summiting the first of the two peaks, light rain began falling and I soon began an 1,100-foot descent down wet, slick, rocky steep terrain, which was very slow going. When I finally reached the bottom near Galehead Hut, the trail was still boulder strewn requiring a slow pace and careful footing on each step. It was getting much cooler, and I had been taking off and putting back on my rain jacket all afternoon, depending on whether I was climbing or descending. Eventually, I arrived at the start of the ascent up to Garfield Mountain, where I would shelter for the evening. On the way up, I encountered a hiker with his dog, and we stopped to talk for a minute or two. Alan and his dog Bowie were from New York and were both drenched from the constant afternoon barrage of rain. He kindly offered for me to go past him, saying he did not want to slow me down. By now the rain picked up and it was getting very cold. My rain jacket was now soaked through to my skin.

The final part of the trail up Mt. Garfield was essentially a near-vertical rock waterfall, which I had to climb for about a quarter mile. The bedrock here was Jurassic age Mt. Garfield porphyritic quartz syenite, which in simpler terms means these rocks are similar to granite but contain little or no quartz. At one point, I thought I had passed the entrance trail to Garfield and would have to turn back, and climb down this precipitous cascade, something I really didn't want to do. Fortunately, I soon arrived at a terrace where a sign appeared indicating Garfield Ridge Shelter was right up a blue-blaze trail. It was almost 4 PM, and the cold rain was now steady with a strong wind behind it. I stopped at a spring on the way up to the entrance trail to refill my water bottles. While there, Fenway caught up to me and we arrived at our evening sanctuary together. There was another hiker already inside when we arrived, Lionheart from Connecticut, and he was reading a book up in the loft area. Not long after did a fourth person arrive. It was Alan and his canine companion, Bowie. There was plenty of room for everyone, and we all hung our wet clothes and gear on hooks along the far wall.

It was a relief to be finally under cover, and we all had friendly chats while we unpacked and got our cook stoves fired up for dinner. I had been starving all afternoon and so I made two dinners, beef stew and later, chicken and rice. It had been a very long day! The structure we were staying in was quite new by AT standards, built in 2011. It had a great design with a loft on top where I spread out my pad and sleeping bag next to Lionheart. Fenway and Alan slept on the bottom, and that night both Alan and his wet dog Bowie shared a sleeping bag. Later, a fifth hiker arrived. We invited him in, but he decided to set up his tent in the rain somewhere, instead. All night long, the relentless wind and cold rain continued to rage into the opening of our shelter.

Day 109
14.5 miles

Day 110 October 8, 2020 Thursday
Garfield Ridge Shelter, NH to Franconia Notch, NH

I got up at 5:30 in Garfield Shelter. It was a very cold night, with heavy rain and high winds all night and into the morning. I made a hot breakfast under cover with my other comrades and was on trail with Fenway by 7:30 am. Strong winds accompanied by a persistent cold rain followed us all morning on our way up to Mt. Lafayette. Despite the bitter morning cold, I was working up a sweat climbing the steep rocky trails, so I took off my puffy coat and rain jacket, just wearing fleece and woollies. Even with both jackets off, I continued to perspire on our approach to the alpine zone. We were both eager and excited to see what lay ahead. Just before the end of the tree line, I put back on my puffy coat, rain jacket and Gore-Tex mittens to prepare for the next part of our journey.

When I emerged from the thick krummholz zone, I immediately felt the harsh sting of the icy wind hit me from the side, and I caught my first glimpse of what was in store for us. Above the tops of the frosted dwarf-spruce trees, all I could see through the blur of blowing snow was shimmering icy rock in every direction. Looking up the mountain, I could see the first false summit of Lafayette above us. We were being assaulted by horizontal snow and ice and 60-mile-per-hour winds. I suddenly began wondering what on Earth I was doing here. Fenway and I looked at each other for assurance and in question of whether or not to proceed, finally conveying nonverbally a resounding, "Lets do it!"

We began our ascent onto the gigantic slabs of Mt. Lafayette granite, with every surface covered in a glaze of slippery ice. Every move of the hand or foot was carefully calculated, making it a very slow process. I was being over cautious so as to not lose footing and slip or slide. It was hard to see the white blazes, now obscured by the raging ice storm and coatings on the rock faces. I was moving excruciatingly slowly, sometimes crawling up rocks. It was a toilsome exercise to stay balanced with the high wind and my top-heavy pack. Fenway was somewhere ahead navigating the same set of seemingly impossible obstacles. The fierce winds would eventually tear away his pack cover, sending it sailing across the mountain, never to be seen again. In addition, the gale force winds eventually would shred his Frogtog rain pants so that all that remained was the linen under-layer beneath the now-absent rubber coating.

After about 90 minutes, and what seemed at the time an eternity fighting my way to the summit, I came to a sign for a cross trail, completely ice covered. The frosted crust obscured the cryptic message beneath, preventing me from reading it. Here the summit elevation was 5,260', and we were high up in the sky in a dense, fierce cloud. Several times I attempted to capture this madness digitally, but my iPhone blinked off instantaneously each time it was exposed to the harsh, frigid wind. For the next 20 minutes, I hunted for just a morsel of indication of where the southbound trail might be, as this was not obvious while I scouted around massive boulders to find my route in this endless blizzard. On one of my attempts to circumvent those icy behemoths standing in my way, I spotted someone with a yellow hat coming up from the south. I waited for him to reappear so I could see the trail and follow his footsteps. He never showed again so I chose a narrow path to the left—presumably the easiest route—and carefully followed it along a narrow rock ledge. There was the man with the yellow hat, with three other hikers taking a break in the protected wind shadow of the cliffs. I maneuvered past them into the blowing snow and continued toward Mt. Lincoln.

Fenway had long since moved onward. From time to time, I was able to use his Altra footprints for some guidance. Overall, it was a very formidable and challenging scramble to the next peak, with all surfaces covered in a smooth layer of ice. I could secure no traction, having to seek out alternate routes either in gravel or vegetation. At times I had no other option than to just scurry onto the ice-covered boulders and take care not to slide off. Eventually I made it to the Franconia Ridge past Mt. Lincoln, and by then, the snow and ice storms had dissipated. The wind was still persisting but in a less menacing way, and with the newfound visibility I could see hordes of day-hikers approaching along the endless ridge trail. Seeing others coming my way gave me a sense of comfort and reassurance. I continued south on the footpath with occasional glimpses of the majestic montane countryside surrounding me. The trail went up and over the last peak at Mt. Haystack, and from there the craggy trail sharply descended, eventually below tree line again, but did not lessen in its extreme grade or condition. I followed the canopied-ridgeline trace for another mile before coming to the Liberty Spring Trail intersection. Here I started the endlessly abrupt climb down—finally arriving at Franconia Notch hours later.

When I arrived at the road in the notch, about 5 PM, I received a text from Fenway saying to call shuttle driver Marlene, and that she would meet me at exit 34A. I checked my map to locate our rendezvous point and began walking along the road. Soon, a New Hampshire state policeman suddenly pulled in front of me with his lights flashing. He dutifully got out of the car and asked me what I was doing. He said I couldn't walk on the highway, gave me a COVID mask, and told me to get into the car. He opened the back right door, and I tossed my pack onto the seat and slid in next to it. I instantly noticed there were no door or window handles on the door as it was shut behind me. He returned to his driver's seat and asked me for my name address and phone number. After calling that in and apparently being satisfied that I wasn't a serial killer, I told him I was a retired geologist from Pittsburgh out here hiking the AT. Looking through the rear-view mirror, he replied, saying he attended Pitt several years ago and even took a geology class. We had a lot to talk about after that, and then proceeded to drive me to the meetup spot with Marlene. He was a good man just doing his job—he could easily have given me a citation but did not. All was well in the end!

Marlene arrived about 5:30 in her mini-van to take me into town. She picked up Fenway only two hours ago and said that he was at a bar in North Woodstock having a beer and a burger, so I requested that she drive me there. I walked into the Pemi Public House, noticed Fenway sitting at the bar talking to someone on his left. The seat was empty to his right so I quietly sat down there and ordered a local IPA. When it arrived, I tapped him on the shoulder and surprised him. We clanked glasses and began interrogating each other about the past 12 hours on the mountain. It was great to talk about our day's experience over a cold, New England IPA.

Later, Marlene returned to pick us up, and drove us to the Notch Hostel—nice place—and we stayed there for one evening in the bunkroom with two others. It was a popular place for hikers and was booked solid for the following several days. One of the bunkmates was a lean and fit looking gentleman, a bit older than me, who recently completed what is known as "The Grid". It turns out many New Hampshire hikers do this, or at the very least know all about it. Completing "The Grid" means that you summit each of the forty-eight 4,000-foot New Hampshire peaks in each and every month of the year. That comes out to the equivalent of summiting 576 peaks, and of course being that this is New Hampshire, includes experiencing many brutally inclement weather experiences—not something for my bucket list.

Day 110
10.0 miles

Day 111 October 9, 2020 Friday
Franconia Notch, NH to Franconia Notch, NH

Today was going to be a ZERO day at Mt. Liberty Lodging, in Lincoln, New Hampshire. This morning we arranged for Mike to take us to the lodge that he and his wife, a past AT thru-hiker, owned and operated. Our plan was to stay there all day and recover from yesterday's ordeal. It was a very pleasant motel overlooking the Franconia Range that we had just conquered the day before. From the parking lot we could see that Mt. Lafayette was still covered in snow. I checked in to my room, and as I was unloading my pack, I could see through the open curtains a beautiful mountain stream flowing right behind the lodge.

After a leisurely morning, Fenway and I shuttled into town to get supplies and look for a good place to eat. Lincoln, New Hampshire was built on the confluence of the "Pemi," or Pemigewasset River and its East Branch, both draining a large portion of the massive White Mountains. We decided to have lunch at a local pizza restaurant, where I soon met Fenway's blind date, Emily. She soon arrived and joined us at our table. We all introduced each other and the three of us talked for a few minutes before I finally left them alone. Emily was quite amicable and apparently a very talented photographer. Fenway had seen her nature photographs on twitter, prompting him to meet her. She later told him that she might join us for our Mt. Moosilauke hike on Sunday. When I heard about this, my first thought was, could she keep up with us? I would soon learn that indeed she could—with little effort!

It was a beautiful autumn day, the weather was perfect, and I really hated to be not hiking today. However, I really needed the rest for my hip and feet. Today's schedule was simply to rest and then plan for a slack pack of 16 miles over the Kinsman peaks tomorrow. Later, we ordered dinner from a local Mexican restaurant, and our favorite shuttle driver, Marlene, delivered it to us.

Day 111
0 miles ZERO Day

Day 112 October 10, 2020 Saturday
Franconia Notch, NH to Kinsman Notch, NH

5:30 AM wake up at Mt. Liberty Lodge. I packed the night before for today's slack packing of the Kinsman tract. Fenway and I met Mike, owner/shuttle driver, outside the lodge office at 7 AM. He dropped us off at a parking lot about 10 minutes away and we were on the trail by 7:30 AM. Today our hike would take us over the gorgeous Kinsman Quartz Monzonite, a white, granite-like igneous rock locally containing an abundance of garnet crystals.

We hit the trail hard and fast, with only a few very short breaks throughout the day. There were several nice views from two of the peaks,

South Kinsman at 4,358' and North Kinsman at 4,293'. Coming off the summits, there were of course many extreme drops and treacherous descents, but we kept up a good 30-minute-per-mile pace over most of the day. Since this was Columbus Day weekend, and a popular foot tract, we encountered many day-hikers and tourists all day long. While boulder hopping across one of the lively mountain streams, I inadvertently stepped on a rock that was not firmly set in the riverbed. As it tilted, my feet were sent splashing into the cold water. I quickly reacted by scurrying over to the dry bank but ended up with wet hiking shoes for the remainder of the day. Just ahead of me, Fenway was maintaining a slightly faster pace with which I was not quite able to keep up.

Later that afternoon when I was within about one hour of our pickup spot, I called Mike to let him know we would be arriving about an hour earlier than originally planned. I then descended down the steep final trail into Kinsman's Notch and eventually made it to the parking lot by 4 PM. There I found Fenway sitting in the grass next to a large rock waiting for me. We were both very pleased to have finished in less time than planned. Very soon Mike arrived in his pickup truck with our 40-pound backpacks to replace the light daypacks we'd been using all day. We said our goodbyes and thanked him once again. We then headed south at the trailhead, looking for a convenient campsite for the night.

The forecast called for very heavy rains tonight and we could already hear the wind beginning to whistle through the trees overhead as dark storm clouds moved in. We chose a stealth site right on the other side of the footbridge over Beaver Brook, about 1/10 of a mile from the parking lot. There was another group of four young people scouting out a second stealth camp spot about 100 yards up the trail from ours. After getting our tents set up, Fenway and I made dinner together down by a dry creek bed, watching the sun setting behind the ominously dark rolling clouds over Beaver Brook Pond. Not long after that, I was in my tent going through my evening ritual of reviewing the day, summarizing of today's events into my spiral notebook journal and planning for tomorrow. All the while, the wind was beginning a series of slow, repeating cycles, where the intensity would steadily build, eventually reaching such a crescendo that I thought my tent was going to be ripped apart and blown away. This was soon followed by equally intense rain, which would continue for the next six hours, well into the very early morning. The oncoming freight train sound of the wind and rain kept me awake most of the night, waiting for what I thought might be the inevitable.

Several hours into the storm, I noticed an array of headlamps flickering all around outside of my tent, generally moving in a direction towards the parking lot. This crazy light show was accompanied by the muffled sound of excited, chaotic chatter. I can only imagine what might have been going on for these four young campers as the storm battered and violated the faux protection of their meagerly assembled shelters.

Eventually the storm subsided, and the wind and rain came to a stop. I may have had an hour of sleep before the familiar constellation sound of my alarm woke me for the next challenging day ahead: Mt. Moosilauke.

Day 112
16.3 miles

Day 113 October 11, 2020 Sunday
Kinsman Notch, NH to Wachipauka Pond campsite, NH

 4:30 AM wake up by my cell phone alarm. I have this annoying habit of getting the best sleep of the night just before my alarm goes off each morning. Most of the night, I was awake listening to intense loud wind, like a freight train building up to a crescendo, then slowly subsiding, repeating over and over again. I didn't put in my earplugs because it was just so mesmerizing hearing the pulsating cycles of the wind and the rain outside. The incredibly intense rain, accompanied by the wind and constant lightning, was just too captivating to ignore with closed eyes and ears. All this to say, morning was not welcome! I managed to get moving in my tent by about 5 AM, organizing things to be repacked, as always, for the day's hike. It appeared that I would not be enjoying my usual hot breakfast this morning as there was too much wind for a flame on my stove. I was imagining this morning's hike with both trepidation and anxiety. Ever since I read Bill Bryson's "A Walk in the Woods," I held a high level of both fear and respect for the word "Moosilauke." And here it finally was.

 Today my morning hike would be Beaver Brook Trail, the main route to the summit. The rocks here are similar to what we had hiked over in the first couple days in the Whites, Devonian Littleton metamorphics. However, here the rock types primarily include mica, quartz, and garnet schists. The trail closely followed the swollen cascade of last night's rain, careening down a steep narrow ravine, for the first 1.3 miles of the steep ascent up the mountain. This morning's hike really turned out to be as craggy, perilous, and challenging as I had expected. After all, I was still in New Hampshire, and there was even a sign posted at the start warning inexperienced hikers to turn back and seek a different route to the top. Much of this route consisted of crazily inclined exposures of near-vertical metamorphic bedrock that no one could possibly hold onto, were it not for the bolted-in steps or rebar added for assistance. To make this even more thrilling, all this was covered in decaying leaves drenched from last night's monsoon-like rain. A slip from one of the steps or footholds could have sent me tumbling down into the roaring torrent of Beaver Brook Falls, just a few feet away to my right.

 Fenway stayed back at camp waiting for his friend Emily to arrive, so I was on my own for the first several miles. About 2 1/2 miles into the trek, I stopped to put on my windbreaker as those familiar gusts of cool air began to intensify telling me the alpine zone was nearing. Or, more likely, I was nearing one of the half dozen or so false summits preceding the top. Just then, I heard more familiar sounds, the voices of Fenway and Emily as they caught up to me. From that point, we all hiked together for the next several hours. Emily brought her camera and was taking pictures all along the way. It was fun having her with us to aptly record this day in beautiful photographs.

We summited Moosilauke at about 10:30 AM, where the whole top of this mountain was enveloped in a slowly rising dense cloud accompanied by the chill of high winds. There was also a freezing mist all around us, which resulted in the coating of every spruce branch in a white layer of frost. This beauty was all around and put us into a jolly holiday spirit. The fog and cold slowly yielded to the penetrating morning sun illuminating the frost-glistening evergreens with a shimmering glow. It was a beautiful day on top of Moosilauke, with a lot of hikers and tourists. Everyone we talked to seemed to exaggerate the extremeness of this alpine environment. Compared to Lafayette, this was a breeze! At the summit, it was mostly foggy and difficult to see much past the trees covering the sides of our mountain. We stayed there about 30 minutes, peering off from time to time in whatever direction the clouds opened up to reveal the scenery. We asked Emily to kindly take some pictures of us at the summit, where the sign showed the location of the highest point. From there, we continued south along the trail following the ridgeline, eventually returning below the tree line where our path changed into a narrow corridor surrounded by frosty, dwarf spruce trees. The descent was far more gentle, stony and abrupt for a short while at first, then quickly yielding to a gradual easy grade for the next 4 1/2 miles to the bottom.

Fenway and Emily went on to have lunch at Jeffers Run Shelter, but I stopped at one of the sunny meadows on the way down and had a relaxing lunch break there. After feasting on a pepperoni and cheese wrap, an assortment of energy bars, and lemon tea, I stretched out in the soft grass to bask in the afternoon heat of that long-absent prodigal sun. Over the course of two hours, the day transformed from a brisk wintery morning to a sunny summery afternoon. I then realized just how much I missed this kind of weather and wondered when I would see and feel it again. Later, I ran into Emily on her way back to meet a shuttle ride from our old friend Marlene. We chatted for a few minutes, took a selfie together, and said our goodbyes. Not surprisingly, Emily, or Mona Leafa as she soon became known, would go on to complete the AT in 2021.

I continued south, and soon met up again with Fenway. We had in mind a stealth campsite on Lake Wachipaucha, just a few miles ahead. After another mile or so, we eventually found the blue-blaze trail that led down to the lake where we found the perfect campsite. The time was about 4:15 PM and this beautiful camp spot was located on the grassy edge of a broad flat promontory overlooking the lake. I located my tent so that it would face east toward the shoreline with open water just a few feet away, and toward tomorrow morning's sunrise.

While we were setting up camp, I noticed a canoe a few hundred yards offshore. I watched for a while as the two fly fishermen carefully cast their lines into the water, over and over again. I used to fly fish a lot in my younger days, and I have to say that their technique was very impressive. As they came closer, I called out asking how the fish were biting, and for the next several minutes, we exchanged comments related to trout fishing.

After today we would be leaving the great White Mountains, following nine amazing days. In a few more days we would be in Vermont, where the long-awaited Green Mountains would stand before us. The remaining trail through western New Hampshire would actually pass through more WMNF lands, but gone would be the 4,000-to-6,000-foot peaks. From here to the Connecticut River, only 2,000-to-3,000-foot peaks remained. Tomorrow was going to be a big day—we planned to shoot for 18 miles of the 42 total miles to finish New Hampshire. I really needed a good night sleep tonight and was hoping for great weather in the morning.

After finishing my journal entries for the day, I looked out my tent and noticed the dark sky was suddenly full of bright, twinkling stars reflecting on the glassy surface of my private pond.

Day 113
11.4 miles
41.9 miles to go until Vermont!

Chapter Twenty
Leaving New Hampshire

Day 114 October 12, 2020 Monday, Columbus Day
Wachipauka Pond campsite, NH to Firewarden's Cabin, NH

    I woke up about 4:50 AM, with earplugs in and did not hear the sound of my alarm. I could still see the stars and a harvest moon in this clear and cold early morning sky. The temperature was 40° and it was looking like the start of another beautiful day. I had a hot breakfast and coffee, packed everything up, and was on trail solo with my headlamp by 6 AM. Leaving camp, I heard Fenway say he would catch up with me in a little while.

    The exit trail out was much clearer and easier to follow than the entrance trail we took to get in. It was an excellent morning hike, the trail winding for several hours through wooded lowlands, but sparsely marked and difficult to follow at times. I arrived at Mt. Cube sometime before my lunch break. The name Mt. Cube turns out to be as metamorphosed as the rocks that make up the mountain. Originally known as Mt. Cuba, its name eventually transformed to "Cube." Its rocks, beginning as lower Silurian sandstones, roughly equivalent to the ridge-forming units such as the Tuscarora or Clinch sandstones of the Appalachian basin, were metamorphosed into a dense, resistant quartzite member of the Clough Formation. The entire mountain is underlain by these tough, glacially smoothed quartzites, creating a beautiful, but slick-surfaced outcrop exposed over most of the summit. While there, I ran into two groups of people who each had been turned around by the confusing signage and somewhat random spacing of some of the blazes. At one of the overlooks, I met a family with two daughters between the ages of six and ten. The oldest was very eager to learn all about what in the world I was doing hiking for 2,200 miles on this Appalachian Trail. I relayed my enthusiastic explanation to her and her family, feeling confident that I had convinced this young lady that an AT thru-hike is definitely in her future.

    I took a lot of very short breaks all morning but never really rested my legs. After 14 miles, and well into the afternoon, I finally sat down to have some lunch. The spot I picked was Eastman Ledges, exhibiting more quartzite outcrops with great views across Jacobs Brook valley to the southwest. I spent the next 40 minutes resting, having my lunch, checking the weather, and sending texts and pictures home. Gazing across the valley to the farthest visible ridgeline, I could see a very faint, but discernable protrusion. This was the fire tower that we were shooting for tonight. It seemed so far away!

    About two hours after lunch, I met up with Fenway again, and together we hiked up to the top of Smarts Mountain. Despite the path following the mountain's rocky backbone, it was a surprisingly easy climb, and we arrived at the cabin near the fire tower in less than two hours. The weather seemed to be suddenly changing for the worse, and it had become very damp, cold, and windy on top of the mountain. Rather than setting up our tents, we decided to camp for the night inside the old Firewarden's Cabin. Rain was predicted for all day tomorrow, and we thought it wise to try to stay as warm and dry as we could for tonight. Several other hikers camped outside in hammocks, and they frequently came inside to visit us in the shelter of the cabin. I was surprised that there was a door and even glass windows. I could even see where an old wood stove and its little chimney used to be, but these installations had long since been removed. There was plenty of room for several more people to spread out on the floor, but only three of us decided to stay inside for the night

In 1939, the Civilian Conservation Corps built the cabin and fire tower after a devastating hurricane season leveled over half a million acres of nearby forestlands in New England the prior year. The cabin was in use until 1973, and then was left unmaintained until several groups (White Mountain NF, HistoriCorps, Dartmouth Outing Club and Green Woodlands) partnered to restore the old structure in 2016.

With all of the structures and campsites here, I expected to find a good water source, and so I didn't fill up at the bottom of the mountain where there were springs and streams. I searched for a running spring somewhere on the mountain, but all I could find were small puddles of murky water. It appeared I would be having plain tuna and uncooked ramen tonight. One of the other hikers had a gravity-filter system with which he had actually produced a copious amount of seemingly clear, pure water from the sludge I had seen. He kindly offered me a cup for cooking my noodles. I gratefully accepted and proceeded to cook my dinner. Later, I returned the favor by offering one of my remaining Almond Joy bars to the benevolent water provider.

After dinner, I checked the weather again and it looked like rain was forecast for the entire day tomorrow. I considered staying here, on top of the mountain, with no water, in a cold cabin for the day. It just didn't feel right. I decided to wait until morning to make the final decision to stay or continue my trekking in this foul weather.

Day 114
18 miles
23.9 miles to go until Vermont!

Day 115 October 13, 2020 Thursday
Firewarden's Cabin, NH to Trapper John Shelter, NH

I woke up around 5:30 am in Firewarden's Cabin on top of Smarts Mountain to heavy wind and rain. Last night there were three of us in the cabin for the night: Fenway, Yurt Guy, and me. I never caught his name, but Yurt Guy was with the group of six other adventurers who were camped outside in hammocks. Last night was very cold, damp, and windy, and I was delighted to have been inside. This morning was exceptionally dreary and bleak. Most of the outside campers came inside one by one to warm up, make breakfast, or just visit. Nearly everyone was discussing the possibility of taking a ZERO day and staying in the protection of the cabin while the storm passed. Also, there were several people coughing and sneezing, and I really did not want to remain there and catch cold from them. I finally decided that I needed to continue moving forward. I packed up, had a quick bite to eat—the only water I had left to make my cold oatmeal and coffee was less than a half a liter of propel energy mix from yesterday—and I was on the trail by 8:45. Although it was quite cold and blustery, the rain was light—barely more than a mist.

The hike down the mountain was a lot steeper and more difficult than yesterday's ascent, but steadily improved as I moved farther down the mountain. I was delighted to feel the air warming as I descended, turning this stormy morning into a pleasant day. Even the leaves on the trees surrounding me seemed to give off a brilliant yellow glow, making the overcast day seem brighter. When I reached the valley floor, I was finally able to stop to refill my water at a clear mountain stream. The trail at the base of the mountain went past Dartmouth Skiway through uncut high weeds, which would have completely saturated me had I not been wearing Gore-Tex rain pants. My destination for the afternoon was Trapper John Shelter, partway up the mountain on which the ski slopes were located.

I arrived at Trapper John's about 3 o'clock. There, I found a modest wooden structure with a large stone chimney fireplace constructed about 10 feet from the open front. Early afternoon heavy rains had left me completely soaked. I climbed into my dry refuge to escape the rain and quickly changed into some dry clothes. Everything wet I hung on hooks to dry. I then made some hot coffee and devoured the remainder of my lunch. I unpacked, set up my air mattress and sleeping bag and crawled in to get warm. After resting for an hour in the warmth of my goose-down bag, I got up feeling much better and later made some lasagna and noodles for dinner. Afterwards, I called the hostel at Hanover Adventure Tours and spoke to Calamity Jane. She cheerfully reserved a spot for me and arranged for my pickup tomorrow in Hanover, New Hampshire. My plan for the next morning was to hit the trail by 6 AM, hike 17 miles to Hanover, and by the end of the day, cross the Connecticut River and be finally—in Vermont.

Day 115
6.7 miles

Day 116 October 14, 2020 Wednesday
Trapper John Shelter, NH to Connecticut River, VT-NH State Line

4:30 AM wake up in Trapper John Shelter. I got packed up, had a hot breakfast, and was on trail by 6 AM. Most of the clothing that I had hung up to dry yesterday remained damp, but it was what I had to wear, and so I put it on anyway. With all the rain I experienced since my June start, I had been doing this wet clothing routine fairly regularly. However, now it was mid-October in New Hampshire, and donning wet clothes in the cold morning was beginning to get old fast! Fortunately, the rain had stopped, and it was actually a quite nice morning. I started out in the dark once again wearing my headlamp, following the trail on Dartmouth Ski Hill to Holtz Ledge. When I arrived at the top, I inadvertently took a trail to the left along the ridge. All along the ridgeline was chain-link fence that was installed in order to protect peregrine falcons. The path eventually led me to the top of the ski hill. I stopped, and through the morning mist I could see the top of a chairlift, and I knew I had taken a wrong turn. I then backtracked, and in a few minutes, was once again heading south following the white blazes.

Everything below me appeared shrouded in fog from yesterday's rain. I followed the trail to the bottom, crossed the valley floor, and as the fog lifted, I got to see a beautiful sunrise reflecting on the mirror-like surface of a beaver pond. Since my arrival to Maine, there had been beaver ponds everywhere. Many were active ponds where I could see a lodge constructed somewhere in the impoundment beyond a carefully constructed dam. Often, century-old dead trees were still standing amongst the grassy flats of a once thriving pond ecosystem now transformed to a grassy meadow. These meadows provided an important habitat for birds and other animals of the forest. After crossing Goose Pond Road, I started the slow clamber up to Moose Mountain. It was a long and steep uphill climb, but not so bad compared to what I had become accustomed to in this state. I arrived at the summit of the North Peak in about 90 minutes, following the trail along the rocky, rising backbone of Moose Mountain. Once on top, the trail stayed flat for a long way as it wound through the wooded highlands toward the South Peak. I didn't see any other hikers up to this point all day, and it seemed like I was the only one out here. I passed by one of the shelters and took a short break on a wooden bench overlooking the valley below.

Coming down, the trail followed an easy grade for a nice descent, bringing me to the western wooded lowlands. I began to see occasional houses, cemeteries, and other signs of culture indicating I was getting close to civilization. I stopped in the early afternoon to have lunch, resting against my backpack next to a large tree, where the warm sun was peering through the light canopy. Following a short rest, I continued southbound to what would be the last bit of a challenge for New Hampshire: Velvet Rocks. From this point forward, it was just a gentle downhill trek for the next hour, through woods and fields. I soon emerged from the woods onto Dartmouth's athletic-field complex where there were students and athletes out jogging around. What a different environment this was compared to where I had just been. Earlier, someone I met on the trail had told me the best place to do a resupply in Hanover was the Co-op. I could see it across the street and headed in that direction to do my resupply. Two-hundred-twelve dollars later, two things came to mind: 1) I had bought less than a week's worth of high-priced, albeit somewhat gourmet, rations; and 2) this was the same price that Rumble from Stanimal's had paid for his sailboat! This certainly was not like re-supplying at the dollar store!

I called Calamity Jane when I emerged from the Co-op to arrange for a shuttle and waited about 15 minutes until my driver appeared. We drove through Dartmouth's campus across the Connecticut River, and I arrived at the hostel at 5 PM. After checking in, I picked up my pack, turned around and there was Fenway. It was good to see a familiar face and we walked back to the bunkroom talking about where we had been over the past couple days. Another hiker I had met in Virginia at Stanimal's was also staying here. Water Queen and her daughter had been hiking together NOBO back in August when I first met her, and now she was flip-flopped and heading south like us. I showered, had supper, did laundry and was in my bunk by 9 PM.

Day 116
16.7 miles

Chapter Twenty-One
The Green Mountains and the Long Trail

Day 117 October 15 Thursday
Connecticut River, VT-NH State Line to Cloudland Road, VT
It was a 6 AM wake up at hostel. I packed everything for the day then made two eggs and four pancakes for breakfast with coffee. Water Queen joined me for breakfast and kindly offered to share some fresh squeezed OJ. She was slack packing today, being dropped off 20 miles south of here and hiking back. After breakfast, we both left with Calamity Jane in the shuttle at 8 AM. I was dropped off at Connecticut River Bridge, and began hiking on Main Street toward Elm, and then uphill on Elm to the AT trailhead. There were a lot of quaint, nicely restored homes along Elm Street in Norwich, a classic New England town. I met several morning walkers along the trail just outside of town. It was an easy grade for the first several miles; however, my pack was very heavy this morning because of my recent resupply. I had to tighten the hip strap and shoulder straps every five to seven minutes all day long; the heavier the pack, the more often I had to do this. With straps tight and snug, the weight didn't bother me. I wrestled with this all day, and hoped as I consumed more food, the problem would lessen.

I made excellent time in the morning, following white blazes up and over many gently sloping hills and down into a deep valley where I crossed under I–84, over some RR tracks, and through a small town on the White River. By 11:30, I crossed over the river, climbed the next hill to the trailhead, and I was back into the woods. On the way up the mountain, I ran into Water Queen. We chatted for a few minutes and told me that Calamity Jane had dropped her off two miles south of the intended start point (Pomfret Road instead of Cloudland Road). Today she was slack packing and therefore carrying a light load. She had recommended that I stop for lunch at the top of the hill where there was a bench with a great view. I also recall telling her about how I was constantly adjusting my shoulder and hip straps, and she commented that another backpacker she knew had always tied his hip straps in a square knot. I initially dismissed this recommendation but tried it several days later and it worked great!

I stopped at the bench on the hilltop to eat lunch. By now I had hiked for almost 10.5 miles and needed a break. As Water Queen said, this spot had a great view, and it was a beautiful warm day. After lunch, I immediately started hiking again, without a break. My pace seemed to slow a bit from not having rested and I knew it was going to be a long afternoon. That five-minute power nap that I missed is always a great way to recharge for the afternoon! When will I learn? For the next three hours, there were long endless uphills on the way to Thistle Hill Shelter. While slogging up the hill, I ran into Lucky Laura, who was an

AT thru-hiker in 1994. She also completed "The Grid," every good New Hampshire hiker's holy grail. Laura was very upbeat and friendly, and genuinely helped me change my attitude for the remainder of the day. She told me where to expect to find water ahead, and where some nice stealth sites would be just past Cloudland. With her advice, I decided to hike to Cloudland Road to get water and then hike up the hill until I located a place to stop and set up my tent for the night.

About a quarter mile past the road, I found a nice flat tent-spot in the woods, cleared the leaves, and laid out logs as a site boundary. It was a beautiful evening when I came across it around 5:15, and the surrounding maples and birches were aglow with color. This would be a nice stealth site, away from the trail that others in a pinch might make use of in the future. After setting up my tent and organizing everything, I fired up my Pocket Rocket stove and made dinner. Tonight, it was ramen with seaweed, sesame sticks, tuna, and shiitake soup mix. One of the best-tasting soups I had made on the trail! I would use the seaweed and sesame again with ramen or rice noodles I bought at the Co-op. I fleetingly imagined that $212 had been wisely spent. I was in my tent by 7:45, a late start for my evening review, planning and documentation. The weather app on my phone predicted rain beginning at 3 AM.

Day 117
16.5 miles

Day 118 October 16 Friday
Cloudland Road, VT to Stony Brook Shelter, VT

At 4:30 AM, I woke up to stars twinkling above in the predawn sky. By 5:30, clouds moved in, and a light rain began falling. I packed up, made hot breakfast in the light drizzle, and I was on the trail by 6:10 AM. I hiked in the dark and drizzle with my headlamp for just under an hour. It was an enjoyable morning despite the rain, as I trudged past countless stone walls next to the paths and roads now occupied by the AT. I also passed by several logging operations where the trail was flagged around some of the more recently cut areas. I wondered why the trees were being harvested. For lumber? Wood pellets? By 11:30, I had my first 10 miles of the day completed, the rain stopped, and I arrived at Winturri Shelter for an hour-long lunch break.

The early afternoon hike took me on a relatively flat and high road following a ridgeline. Up ahead was a signpost indicating a blue-blaze trail to "The Lookout." Although I was in a hurry and didn't want to stop, I recall Shutter recommending that I sojourn here to see this unique cabin. So, I did. There was a ladder on the outside of the cabin leading up to a viewing deck on the roof, with spectacular 360° views of the Green Mountains. For the remainder of my time in Vermont, I would wander in these ancient mountains.

The Greens continue south into Massachusetts, but there they are known as the Berkshires. This mountain chain, although less impressive in its elevations than many others in the east, has a tectonic history which includes three orogenies, or mountain building events: Grenvillian, Taconian, and Acadian. Ancient continental margin deposits were thrust west during the Taconian Orogeny and then again to a lesser extent during the Acadian orogeny. Great plutons of granites and other sialic magmas were emplaced during this time and slowly exhumed by continued uplift and erosion. Most of the rocks seen at the surface in the Green Mountains today were once so deeply buried to have been transformed to metamorphic rocks such as gneisses and schists.

At "The Lookout," I climbed back down the ladder and hopped up onto the front porch to check the cabin door. To my pleasant surprise, it was unlocked. I entered to the faint smell of smoke from the large stone fireplace, combined with fragrant cedar planking covering the inside walls and steep vaulted ceiling. Were the timing right, this would have been a great place to stay for the evening. However, I had a lot of miles to cover today, so I didn't spend much time here and quickly returned to the trail.

Soon after, the rain increased to a heavy downpour, which lasted the remainder of the day. I was still wearing my summer rain jacket, which generally would only keep me dry in light rain for an hour or so. As a result, I was beginning to get very wet and cold. The only way to stay warm was to keep moving at a brisk pace. I had my goal set at 20 miles today and looked forward to changing into dry clothes at Stony Brook Shelter. The hike up to my destination seemed much longer than the 0.6 miles advertised on the signpost. It was remarkably steep including several rebar stairs and ladders. After crossing over the recent rain-swollen slough cascading over the feldspathic quartzite ledges of the Stowe Formation, I finally arrived at the shelter, in heavy rain, about 5:30. I quickly unpacked, changed into dry clothes, set up my air mattress and sleeping bag, and made ramen noodles for dinner. After finishing, I hung my bear bag and was in my sleeping bag for the night. It was a very long day!

Day 118
19.9 miles

Day 119 October 17, 2020 Saturday
Stony Brook Shelter, VT to Churchill Scott Shelter, VT
I woke up late, due to heavy rain throughout the night and continuing into this morning. Gazing around the shelter, I was surrounded by various articles of my clothing hanging by hooks—still wet from yesterday. I remained here for longer than I wished, waiting for the rain to stop. The drab weather outside really provided little inspiration. Eventually I made breakfast and was packed up and ready to go by 10:30 AM, in light rain. As

I was leaving, I noticed patches of snow on many of the north-facing slopes, and as I gained elevation, I found even more snow covering the ground.

It was a steep initial climb to the top of Quimby Mountain, but for most of the remainder of the day, I was hiking over gentle hills and valleys. Since my arrival to the Green Mountain State, the terrain was much less dramatic, and the daily hill-climbs much tamer than the Whites or Mahoosucs. Today my journey led me through state-park lands as I was approaching the merge-point with the Long Trail (LT). I passed by many tourists and day-hikers, due to the trail's proximity to Highway US 4 and Gifford Woods State Park. I also noticed that I had cell service on my phone all day long, and I texted and sent several pictures back home and to friends while at various viewpoints. It had been a cold, wet start this morning after leaving the shelter, but by noon, the clouds began breaking up, revealing the welcome warmth of the sun.

The LT is the nation's oldest long-distance hiking trail stretching 273 miles from the Canadian border to the Massachusetts state line. In 1909, James P. Taylor envisioned the trail, and the work began in 1910 by the Green Mountain Club. Over the years, parcels of land have been acquired along the trail by the Green Mountain Club, and like the AT, nearly the entire footpath is now protected. The AT connects with the LT at Willard Gap, and from this point, the two trails share the same path southward for 105 miles.

I continued my hike until sunset, finally arriving at Churchill Scott Shelter area where I quickly put up my tent and made some dinner. My dinner menu consisted of ramen, seaweed and vegetable soup, and a tortilla wrap with pepperoni and string cheese. I could tell my 4-ounce gas canister was running low. The small 4-ounce size is just not dependable for this type of hike. I would need to go with the 8-ounce cans from now on.

I was in my tent by 8 PM to begin my map review and daily journaling. It was becoming quite chilly with the temperature reading 38°, and tonight's low expected to be 33°. Because of the possibility of freezing temperatures, I slept with my cell phone, charger, and water filter inside my sleeping bag.

Several days ago, when we were at the hostel together in Norwich, Fenway, Water Queen, and I all planned to meet tonight here at Churchill Scott, but it was looking like they weren't going to make it. Tomorrow, I could look forward to better weather and gentler terrain after Killington peak.

Day 119
11.4 miles

Day 120 October 18, 2020 Sunday
Churchill Scott Shelter, VT to Clarendon Shelter, VT

With a 4:30 wake up, I was on trail by 6:15 AM. It was cold oatmeal and hot coffee this morning, as my 4-ounce gas canister was nearly empty.

It was a pleasant, snowy climb to Killington peak. Cooper Lodge, near the peak, was a beautiful stone structure and I stopped in to check it out. As I approached, I could hear voices coming from inside the cabin, and when I opened the door there was Peter, a backpacker making his way south from Canada on the Long Trail with his pug dog, Mr. Bill. We chatted for a few minutes, and I soon left the cabin and resumed my southbound journey. There had been a lot of snow on the mountain last night, and someone—presumably Peter—had made a small snowman just behind the cabin.

Later as I was making my descent down the narrow trail off Killington Peak, I ran into two local Vermont women. We talked for a couple of minutes, and they relayed to me several stories of their previous hikes to Killington peak. One involved unexpected heavy snowfall on an October night while tent camping on the top of the mountain at Pico Camp. They told of how their tents were collapsed by the snow when they arose the next morning. As I continued down the mountain, the trail seemed to serpentine more often than necessary, almost as if it were intentionally built that way to replicate skiing moguls. I was, by the way, still on Killington Mountain.

The terrain was actually very easy coming down, but I still had difficulty making good time. While still on the mountain, I tried to FaceTime my daughter Jessie several times, with no success, as today was her birthday. I spent 10 hours hiking today, but only was able to get in 14.5 miles. My feet were still bothering me, and I probably was holding back a bit because of it. Despite the sore feet, I really enjoyed the remainder of the day and met several other hikers on the trail. The weather was nice, there were lots of picturesque stream crossings, and by 4:30 PM, I arrived at my destination for the night.

Clarendon Lodge was a relatively new and very clean shelter, and someone had stocked it with freshly split firewood, mostly birch. This was an unusual treat—and much appreciated. Next to the lodge was a clear, cold, gurgling stream that could be heard from the camp area. I decided to make a fire, both for cooking dinner, and as a surprise in the event that Fenway and Water Queen later joined me that evening. I hadn't heard from either of them in several days, but with the lower-than-expected miles I was logging each day, I thought they might finally catch up tonight. It turned out that never happened, but it was still great to enjoy the warmth and ambiance of a campfire for a change.

I let the fire burn down, and soon had a nice bed of coals for cooking. There was a grill hanging next to the woodpile, which I placed over the coals, then heated my water to a boil, and made my ramen noodles. I also grilled chorizo sausage, which I then tightly wrapped, along with string cheese, inside of a tortilla shell. More leftover gourmet items from that $212 resupply in Hanover! Finally, I grilled this over the hot coals for a few more minutes, continuously turning it from side to side in order to melt the cheese and crisp the tortilla shell to perfection. The result was a hot grilled burrito, which proved to be a delicious addition to my normal dinner of ramen noodles.

I had set my tent up across the little field from the shelter so that the door was facing the woods in the direction of the nearby brook. After I put out the fire, I watched the now black sky and the myriad sparkling adornments overhead. It was dead silent and peaceful with only a sliver of the waxing moon visible through the trees. I was in my tent by 7:40 PM.

Day 120
14.8 miles

Day 121 October 19, 2020 Monday
Clarendon Shelter, VT to Little Rock Pond Shelter, VT

I awoke at 4:18 AM to a beautiful morning; the stars were out and surprisingly there was no chill in the air. I was on trail with my headlamp by 5:40 AM with the first part of the hike proving to be quite interesting and challenging. The path I was following quickly degenerated into a steep rock pile filling a narrow gorge where I found myself navigating down large boulders with occasional series of rock steps. When I finally got to the bottom, the trail led to highway 103 where, across the street, I could see a trailhead parking lot. I decided to hike out to Loretta's Deli, just 0.6 miles away, to get resupplied for the next several days. I had called the store before leaving the mountain, and the recording said the doors open each day at 6 AM. I made my way up the busy road, and on my right, I eventually saw a sign for the deli. When I arrived, I noticed there were no cars, and the store looked to be closed. Uh-oh. The sign on the door apologetically explained that the new owners would be opening the store soon. I turned around and returned to the trailhead parking lot. Still needing to re-supply, I got back on the trail to continue hiking south for another 6.3 miles to the next highway intersection, where I might be able to hitch a ride into Wallingford.

It was a very lovely day to be on the AT. I passed several ponds and summited Bear Mountain. About a quarter of mile before the next road, I met Ryan standing with his bike next to the footpath. We talked for a while, and I told him about my plan for hitching a ride to the next town for a re-supply. He said he'd gladly give me a ride into town if I was still there when he finished his activities for the day. About 30 minutes later as I was coming back up to the road from getting water, I spotted Ryan pulling out of the parking lot. He saw me, turned off the road, and graciously offered me a ride into town. We had a pleasant chat on the way, and he dropped me off at the Dollar General store downtown. We said our goodbyes and I thanked him and went inside to buy my usual rations for the week. After reloading my food bag and backpack, I was off to find a way back to the trailhead down the highway.

As I was crossing the street, I noticed someone up ahead carrying a guitar. It was Ryan. He had been sitting on the patio of one of the local restaurants playing some tunes and practicing his guitar for the past hour. He offered me a ride back to the trailhead and wanted to hike with me to the top of the next mountain, White Rocks. Ryan seemed like a very nice person, and we agreed to hike together for the next hour or so. He explained that although he enjoyed being in Vermont, he really missed living out west. Ascending the mountain, we talked about some of our past adventures and other places we would like to explore. As usual the trail was steep and rocky, and I was surprised how easily Ryan was able to keep up with my pace wearing a pair of crocs with no socks. At the top we found ourselves amongst a bizarre menagerie of oddly placed quartzite cairns. From there, I continued following the white blazes, while Ryan took off down a blue-blaze trail for White Rocks, to do some boulder scrambling. There, massive slides of white blocks of quartzite following the last ice age litter the steep hill slopes.

It was already mid-afternoon, and the skies were beginning to look dark and stormy. The weather forecast predicted rain by 5 or 6 PM, and I still had another seven miles to get to my target destination. I looked again at my map and realized that there was another shelter on the way at Little Rock Pond. I got there about 4:30 and decided it was enough for the day. There were four other backpackers there, but plenty of room inside, and I had one of the large bunks on top to myself. I met Brent and Laura, a retired couple from New Hampshire who had a son working in Shanghai, China. I told them that I had a son working there too! In addition, they had also hiked through parts of Vietnam, Cambodia and Thailand. I told them that I hiked through parts of each of those countries as well with my son Josh. What a coincidence! We shared a lot of fun stories over the next hour.

The two others in the shelter were Caroline and Topo. Caroline was a thru-hiker from Spartanburg, South Carolina, and she was doing a flip-flop, similar to mine, but was planning to finish her trip in Boiling Springs, rather than Swatara Gap, Pennsylvania. We talked briefly about hiking together for the remainders of our trips. However, after discussing daily miles, it became clear that our average paces didn't match well.

Fenway and Water Queen texted me later that evening. They made it to Greenwall Shelter, about seven miles back near White Rocks. Due to heavy rain in forecast, they decided to get a ride into Manchester Center tomorrow for another ZERO day. I remember thinking it was too bad that Fenway couldn't keep up the same pace we hiked through Maine and New Hampshire when, on many days, he outpaced me. It seemed he had really slowed down for some reason, and it was then that I realized he was likely never going to catch up.

Day 121
13.6 miles

Day 122 October 20, 2020 Tuesday
Little Rock Pond Shelter, VT to Bromley Shelter, VT

I awoke at 4:30 AM on a second-level bunk at Little Rock Pond Shelter. The intermittent rain I heard last night was now barely a drizzle on the tin roof. I tried to get all my gear packed up with minimal disturbance to the others and had breakfast of cold oatmeal and hot coffee on the lower level. I was on the trail by 5:40 AM, with my headlamp shining my way like a beacon in the darkness for the next 45 minutes.

The trail was in pretty good condition this morning, and the rain finally stopped. The path led me down along Beaver Brook for several miles, eventually turning uphill, where I slowly gained elevation toward Baker Peak. The rain returned, and just before the summit, I noticed a sign saying there was an optional bad-weather detour ahead. I decided to just stay on the regular route, and for over a hundred yards, the trail followed a Mesoproterozoic age thinly layered schistose quartzite knife-edge along Baker Peak. It was very steep and slippery, and I carefully moved step-by-step so as not to lose my footing on the slick wet rocks.

Later, the trail also summited Peru and Styles peaks over more 1.3-billion-year-old gneisses and schists, then dropped down to Mad Tom Notch, and back up a gently graded path to Bromley Peak. Bedrock here was also the same age as at the last two peaks. When I reached the top, the trail came out at the top of Bromley Ski Resort, near the terminus of the Sun Mountain Express Chairlift. Bromley is known as "Sun Mountain" because it faces south rather than north, as do most resorts, and so receives more sun.

The AT then followed Run Around, a beginner ski trail, downhill for about a half a mile and then turned abruptly into the woods. Another mile down the hill was Bromley Shelter. Arriving about 4 PM, I decided to settle in there for the night, made dinner, unpacked, and was in my bag soon after. It was a long, rainy day and I never saw any other hikers. I would be looking forward to better weather tomorrow.

Day 122
17.8 miles

Day 123 October 21, 2020 Wednesday
Bromley Shelter, VT to Stealth campsite on Black Brook, VT

The morning started with a 5 AM wake up. With an empty fuel canister, I had a cold breakfast. Last night, after I had already hung my bear bag, I found candy wrappers in my pocket. Rather than going out and retrieving the bear bag just to add the wrappers, I decided to simply double Zip-lock them and leave them on the table under some rocks. This morning, I discovered those sealed tight bags and candy wrappers had been chewed to pieces by mice. As I was packing my backpack and food bag in the dark by the light of my headlamp, several mice were scampering around the shelter looking for food. Later that day, I noticed a small hole chewed into my food bag. There was also a hole in the plastic bag containing cheese sticks, and a hole in the plastic bag containing tortillas. I hadn't noticed this when I brought my bear bag down from the tree. It appeared the damage was done in the 15 minutes or so the mice were scurrying around while I was packing. I patched my bag with some Leuko tape, and it seemed to hold pretty well.

Hiking today was pretty easy. The trails for the most part were less rocky and less rooty than usual. I made good time on several long, flat stretches. This morning was foggy, bleak, and overcast but rain never came. In the afternoon, the sun burned through the clouds and warmed the day into the 70s. The hike up to Stratton Mountain was a gentle incline, three miles up, three miles down, with a fire tower at the summit offering stunning, 360° panoramic views. To say the geology in Vermont is complex would be grossly understated. This mountain, and much of the LT and AT here is along the eastern flank of the Green Mountain Massif, a thrusted anticlinorium containing 450- to 1,300-million-year-old very tired rocks. Parts of the trail up the mountain were extremely wet and muddy with the highest elevations comprising primarily conifers amongst the moss-covered forest floor. I only encountered four hikers all day on Stratton Mountain, which surprised me knowing how popular and accessible this mountain was.

I began thinking about where I would resupply in the next two days. My battery charger was dead, and both my cell phone and headlamp were in desperate need of a recharge. I kept thinking about these things throughout the morning, and during my lunch stop at the Goddard Shelter. I was in survival mode and had to formulate a plan. Later, I found a great tent site next to Black Brook. I refilled my water bottles, washed up in the river, and rinsed out some of my socks. For dinner I made a cold ramen, dried cranberries, and goat cheese concoction, which was actually quite palatable. More good eats from my $212 resupply! Today had been a very long day in miles, but the terrain was easier than usual.

Day 123
20.5 miles

Day 124 October 22, 2020 Thursday
Stealth campsite on Black Brook, VT to Melville Nauheim Shelter, VT

Because I just did a 20-mile day, it took me a bit extra time after my 5 AM wake to get packed and ready to go. Little did I know that I was in for another 20-mile day. The trail started out mucky and full of big boulders, but as it gained elevation toward the ridge top, it became smooth and fast. I could only imagine what kind of conditions these parts of the trail would be like were this the spring rainy season.

It appeared that perfect hiking and weather conditions had finally come together. Weather was cool and sunny all day, and the trail, for most part, was as good as could be expected anywhere. While on the ridge top, I had cell service, which allowed me to text messages to my son's friend, Dave about our planned meeting time on Friday. I was hungry all day. The food I had eaten for snacks and lunch just didn't satisfy me and I was starving. That afternoon, much of my trek was along narrow, flat ridgelines where I was able to delight in nearly continuous glimpses of the surrounding mountains.

When I arrived at Nauheim Shelter, at 6 PM, I had just enough time to set up my tent and get organized before darkness set in. Brent and Laura from three nights ago were tenting here too. It was nice to see them again and they invited me to come by their campsite later. I said I'd come over for a chat after getting everything organized. I later went down to their campsite and had dinner with them. We talked for about an hour and Brent graciously lent me his stove to cook my food since my gas canister was empty. Laura, who is Lebanese and residing in New Hampshire, told me all about their recent trip to her home country. They shared fascinating stories about lands I had never seen. I also told them of my plans to make it to North Adams tomorrow and visit with my son Josh's roommate from college.

I said good night to my new friends and was in my tent by 7:00. Initially, I thought it was only 16 miles to North Adams, but when I reviewed my map that night, it turned out it to be 20. That would make three-20 milers in a row. Ouch!

Day 124
20.0 miles

Day 125 October 23, 2020 Friday
Melville Nauheim Shelter, VT to MA 2, North Adams, MA

4:17 AM wake up and I was on trail in the dark by 6 AM. Earlier, my phone died—just as I opened it to see the time. It either said 1:17 or 4:17, I couldn't be sure. I decided to get up either way since I had a huge day of hiking ahead of me, and if it were indeed 1:17, I would have that much more time to get in my 20 miles. After about a mile and a half, I was nearing Vermont Route 9. The horizon before me began to lighten ever so slightly. Apparently that last flash of life from my cell phone must have read 4:17.

Following an early morning cliff climb over boulders and ledges of Proterozoic-age gneiss to the top of Harmon Hill, it was relatively easy hiking the rest of the day. There were some great views along the ridgeline where the trail was long, straight, smooth, and flat. The term "turnpike" came to mind. The trails in the lowlands were wet and boggy and I was often walking on slick boardwalks for long distances. I was on the lookout and saw evidence for beaver dams throughout the day.

Later I met a local hiker at the Vermont-Massachusetts border who explained to me about the three false summits of Mt. Greylock, which I would be summiting tomorrow. I told him that by now I had come to expect it, but that I appreciated the heads up. I also inquired what time it was since my phone had been dead since 4:17 this morning—he replied, "2:30 PM." I now had a timestamp for leaving Vermont, the Long Trail, and the enchanting Green Mountains.

I was now in Massachusetts, my eleventh state since beginning this fantastic journey. Just past Eph's Lookout on my final descent towards South Adams, there were some nice views and once again an opportunity to take a bad-weather trail around what I expected to be another dreadful "drainage ditch" downhill. Of course, I chose the normal route, and of course it was unreasonably steep and chock full of irregularly positioned boulders. It eventually ended in a Cambrian-age quartzite scree pile at the bottom. And of course, this is what made it so much fun! It wasn't terribly difficult navigating, but the edges of the quartzite blocks were pretty darn sharp, and I had to be careful with every step. For the next mile or so, the trail was very gentle and followed Sherman Brook all the way into town. I passed what appeared to be an old, round concrete reservoir, perhaps once used last century as a freshwater reservoir during dry spells. Soon, the sound of light traffic began overtaking the faint wind rustling through leaves all around me.

I emerged from the woods into South Adams-Greylock at about 4:30 PM. I hiked over to the laundromat to do my laundry. After loading the washing machine, I went to the vending machine, and apparently bought a small box of softener (not soap), dispensed the contents of the box into the washer, and started the wash cycle. The alert young attendant evidently noticed this act of domestic incompetence, and dutifully scurried over with the correct detergent. It was a kind gesture, and he didn't even charge me for my silly mistake. While waiting for the wash cycle to end, I called for a room at the Willow Motel in Williamstown. They had a vacancy and offered to send over a shuttle to pick me up. I next contacted Dave, Josh's roommate at Centre College, to make dinner plans. Dave was now living in South Adams and teaching political science at one of the local universities. After college, Dave and Josh both worked at Yosemite for a season, then traveled the world together with some other friends to places like Mexico, Ecuador, Peru, Samoa, Bosnia and New Zealand.

The shuttle soon arrived, and my driver was a friendly, older gentleman who had recently arrived from Mumbai, India. I sat in the back seat and we both struggled to communicate as he drove me to my motel. I checked in, found my room, unpacked my gear, and showered. Later, Dave came by to pick me up, and we drove over to the Purple Pub, in the heart of Williamstown, on the campus of Williams College. It was a beautiful campus, a great little pub, and the burger and local draught beer were a welcome change from my usual dinner of ramen. I also had a great time hanging out with Dave for that couple of hours. He dropped me off at my motel by about 9 PM.

Day 125
20.0 miles

Chapter Twenty-Two
The Berkshires

Day 126 October 24, 2020 Saturday
MA 2, North Adams, MA to Mark Noepel Lean-to, MA

This morning was a late start, and I had a lot of chores to do. I needed to re-supply at Family Dollar and the food store, search for a fuel canister, and make several calls home. My kind and amicable shuttle driver dropped me off near the AT trailhead around noon. I was finally going to be trekking in the Berkshires. I hiked to the top of Mt. Greylock, following the three false summits I had been expecting. After getting to the top of the first peak, there was a long descent into a deep gap between Prospect Mountain and Mt. Williams. The next two peaks had much milder gaps following them, and eventually I made it to the top of Greylock, where the very resistant Greylock gneiss underlies the peak.

It was a Saturday, and as I was going up, I passed many tourists and day-hikers going in both directions. Most people I encountered were in fall or winter attire, some with hats and gloves. Because I was carrying a heavy pack and mostly gaining elevation, I was still working up a good sweat, even in shorts and a sleeveless shirt, inviting startled looks and inquiries along the way.

By the time I reached the summit, it was very cold, windy, and overcast, and the dense fog afforded no views. I added a long-sleeve shirt and my trusty rain jacket to fend off the bitter cold. With several large parking lots just yards away, this place was crawling with tourists. The white blazes I had been following ended abruptly at the long, circular sidewalk surrounding a large, 100-foot-high War Veteran's Memorial. In the fog it was an eerie sight, appearing like a lighthouse on a foggy shoreline somewhere. It wasn't at all obvious where the trail went from here. I talked to a couple of local mountain bikers, and they pointed me in the right direction. There were a lot of blue- blaze trails heading off to parking lots or other points of interest on top of the mountain, and it required a fair amount of vigilance to stay on the white-blazed AT.

Soon I was well on my way to the next shelter, arriving just before sundown. When I arrived, it appeared there was no one else there, however a pair of trekking poles were leaning up against a side wall. Just then I heard a voice from above me call out, "Hello." In the loft above me was Adam, from Springfield Massachusetts. We talked for a bit, and he mentioned that in the morning he was planning to stop at a Dunkin' Donuts at the bottom of the mountain, in the tiny hamlet of Cheshire. I thought this sounded like a great idea and told him I would join him.

I set up for the night on one of the lower bunks due to the wet and cold, as the temperatures were expected to fall into the low 30s. I think it was another good call! For the next 45 minutes, I reviewed some maps and made some notes, and then it was lights out. Mice were active in the shelter all night, but they didn't bother me one bit.

Day 126
9.6 miles (Half-day only, noon start)

Day 127 October 25, 2020 Sunday
Mark Noepel Lean-to, MA to Kay Wood Lean-to, MA
 I arose at 4:30 in Mark Noepel Shelter. It was very cold in the early morning—somewhere in the low 30s. I quickly got dressed, packed, had some cold coffee and an energy bar, and was on trail by 6 AM. I was looking forward to today because the trail passes through two small towns, Cheshire and Dalton, Massachusetts. Adam got up about the same time and we hiked out together. The entire four miles down to the highway was downhill, and we arrived at Dunkin' Donuts by 8:15 where I ordered three donuts, two sausage and egg bagels, and coffee. What an awesome breakfast! I looked in a few local stores but was unable to find a propane fuel canister. After some discussion, Adam said he had several at home and kindly offered to sell me one of his. When he called his dad in Springfield to pick him up, he also asked him to bring another along. Later, he hid it next to the road near Kay Shelter where I would be staying tonight and sent me a picture of where the canister was hidden so that I could retrieve it.
 While I was in Dalton at Cumberland Farms, eating ice cream out front, I met Jason Phelps, known also by his trail name, Rattler. Two years ago, Jason hiked from Daleville, Virginia to Massachusetts during the winter months. He said he was planning to come up to Kay Shelter later in the day, to be on the trail for the first time since he got off two years ago. I said that was where I planned to be for tonight and that I anticipated meeting him there. We both were looking forward to sharing our experiences and I was particularly interested in hearing about winter hiking on the AT.
 On my way into the shelter, I met a young couple just leaving, who said that Jason was up by the fire waiting for me. The smell of the burning wood was in the air, and I was anxious to warm my hands over the campfire. When I arrived, I became a bit concerned about the wind because of all the dry leaves around. I raked some away so that no sparks would ignite them. It was interesting talking to Jason about his winter trip; he had a lot of interesting stories about getting temporarily lost because of blown-down trees blocking trails, or snow covering the white blazes. He told me that for years now, his job has been making snow at a local ski area, trampling through the snowy slopes all night long. This evidently had uniquely prepared him for a winter backpacking journey. After exchanging AT tales for about an hour, we said goodbye and he headed back to his car. Later, I used the fire for cooking my dinner (even though I now had fuel!).

Before turning in for the night, I extinguished the fire with water from the creek. After I was in my tent for a while, the wind began blowing harder. I was concerned about it reigniting the coals, so I returned to the creek for some more water, and doused them once again. After returning to the tent, lying there listening to the wind a while longer, I felt there was still a possibility of sparks being blown onto the dry leaves. So, I repeated the whole process a third time, but this time with enough water to be sure there would be no more danger of fire. I slept much better afterwards with the confidence of knowing I would not be starting a forest fire.

Day 127
16.7 miles

Day 128 October 26, 2020 Monday
Kay Wood Lean-to, MA to Upper Goose Pond Cabin, MA

I awoke at 4:30, listening to the constant pattering of rain on the roof until 5:30, when I finally got moving. Poor decision! By the time I got organized, packed up, and ate breakfast, it was 8 AM. I had lost two good hours of hiking. The morning was very cold and dreary. It slowly warmed up to the 50s, but remained bleak and damp all day. My feet were also wet and never dried out, even though the rain stopped in the morning.

For most of the day, the trail followed a fairly even grade and there were just a few minor hills to climb. Around noon, I passed by a beaver pond in the middle of which I could see a carefully engineered mound of sticks built as the lodge. Evidence for this crafty rodent was all around, including many partially chewed and downed trees lying all around. A cage had been installed around a culvert, beneath a human-built earthen dam and road, so that the beavers could not access it, and thereby plug it up and flood the road. Devices like this one were invented by Vermont native Skip Lisle and were craftily coined as Beaver Deceivers. These ingenious devices work by allowing the beavers to survive and flourish in their impounded paradise but preventing them from damming up outlets that would intolerably raise water levels for their human companions.

Toward late afternoon, I had to cross over an interstate highway, then climb up and over a hill toward Goose Pond. There was a well-known cabin there, and I really wanted to check it out. I could still hear the sound from the highway even after I crossed over the hill. I arrived at the end of a 0.5-mile blue-blaze trail at Goose Pond Cabin by 5:30. I went over to the cabin to see if I could get inside, but it was locked for the season.

I found a nice camp area nearby, set up my tent, got water and made dinner. As I was setting up my tent, I began feeling familiar twinges of pain in my lower back and worried it could become a real problem. For many years, I had issues with two ruptured lumbar discs. They would sporadically cause me pain to the point of immobility for several days.

From the time I began hiking the AT in early June, I hadn't had a single problem with my lower back, and I sure didn't want to start today! After finishing dinner and putting things away, I carefully got into my sleeping bag, did some reading and journaling, and tried to get some sleep.

Day 128
17.6 miles

Day 129 October 27, 2020 Tuesday
Upper Goose Pond Cabin, MA to Mt. Wilcox South Lean-to, MA
    Today's was a 4:30 wake up, and I was on trail by 7 AM. It was cold overnight—down to the low 40s. To my great relief, my lower back felt fine this morning as I started out the day on moderate to easy terrain coming out of Goose Pond cabin area. I crossed over a small country road, and to my right, was the town of Tyringham, Massachusetts. I continued across into a wide grassy bog and I found myself walking for a long way on an old boardwalk. Many of the boards were broken or tilted, and they were all wet. At one point I slipped on one of the slick boards and my foot ended up in the deep muck. Yuck! I couldn't be sure, but it appeared the bog was the result of beaver-dam construction downstream from my crossing.

    I soon arrived at the entrance to Tyringham Cobble, a nature preserve with lots of hiking trails and scenic vistas. I found a perfect lunch spot on an outcrop overlooking the town of Tyringham, Massachusetts. The cobble, or rocky ledge that formed the overlook, was a thrusted block of Precambrian strata over top of much younger Paleozoic strata. While eating lunch at the overlook, I met two young ladies who were section-hikers from Massachusetts. They eagerly presented me with a preview of some upcoming trail highlights, and then recommended a great burger joint on the way to Great Barrington.

    One of the highlights I was told about was an unnamed notch ahead, described as "Jurassic Park." Later, while hiking through the notch, I could see why. There were sheer cliffs all around covered in moss and ferns looking like something from another geologic time period. Later, I pitched my tent near the Mt. Wilcox South Shelter. It was one of the oldest on the AT, built in 1939. Down the trail about 1,000 feet was a newer shelter, built in 2007. The spring there was almost completely dry, and one of the poorest water sources I encountered on the trail. I was able to collect enough to boil and make dinner, but really didn't want to run it through my filter for the sake of not clogging the membrane.

    I chose to camp out in my tent tonight because originally the forecast was for 50% chance of rain in the morning. After I was set up and preparing my dinner, I checked the forecast again. It now showed a 90% chance of rain through 10 AM tomorrow. "Uh-Oh. Looks like I will be tearing down in the rain," I thought.

I called several hostels in Salisbury CT where I'd be staying the following evening. There were two ladies that ran hostels in that town. One was 90 years old and very friendly but said apologetically that she could no longer accept hikers because of her doctor's orders. This was likely COVID related, and at her age I certainly understood and agreed that she was doing the right thing. I called the other hostel owner, Vanessa Breton, but I only was able to leave a message. I would have to wait until tomorrow to know if I had a place to stay for Thursday night. Either way, I would still need to re-supply when I got to town.

Day 129
15.9 miles

Day 130 October 28, 2020 Wednesday
Mt. Wilcox South Lean-to, MA to The Hemlocks Lean-to, MA

I awoke at 5 AM in my tent to a cold pouring rain. I packed everything up under the cover of my vestibule, then ran my pack over to the shelter, and finally tore down and packed a wet tent after putting it into a plastic bag. It was a long day with a steady rain all morning. Later in the afternoon, it became partly sunny and was much more pleasant. Much of the afternoon hike was on long flat boring stretches across a valley that seemed to go on forever. It reminded me of crossing the Cumberland Valley in Pennsylvania. Ironically, I met a young man from Camp Hill, located in Pennsylvania's Cumberland Valley, now living in Connecticut. After chatting for a minute or two, he said that he was interested in doing an AT thru-hike. I told him all about the trail and gave some recommendations on how to get started. Following the discussion, he seemed really excited about what was in store for him.

Later, I hiked up the mountain to get to Hemlock Shelter—my destination for the day. It was a long way, but in order to get into Connecticut by tomorrow, I had to do this. On my way toward Hemlock, I ran into two hikers who were lost and going the wrong way. I showed them a map and pointed them in the opposite, and correct, direction. They had no flashlights, and no map apparently. It was about 5 PM, nearing sunset, and had I not seen them, they would have been wandering around for a long time in the dark. I might have saved their lives. Ascending the hill to Hemlock Shelter, I passed several nice flowing streams. It was very damp, cold, and wet, and had been a really long day! Because of the foul weather I again decided to stay under cover for the evening. After a hot meal of ramen noodles and PB on a tortilla, I was in my bag on one of the dry bottom bunks by 7 PM.

Day 130
19.8 miles

Day 131 October 29, 2020 Thursday
The Hemlocks Lean-to, MA to CT 41, Salisbury, CT

I woke up at 4:30 in Hemlock shelter, it was about 40° and just misting rain. Yesterday's rain made the air heavy with moisture. My sleeping bag was wet, especially the toe box, which was still soaked from touching the ceiling of my tent the previous night. I made a hot cup of rice noodles for breakfast and then ate two energy bars with my coffee. By 6:15 AM I was back on the trail with my headlamp, hiking up a dark, misty trail to begin my Mt. Everett ascent. Having read about this mountain, I was expecting a significant challenge, but the morning climb was remarkably easy.

After making it over the summit, the rain returned, but it was an easy grade down into the gap, and then up to Mt. Race Ledges. For a mile or more along this open ridgeline, there were awesome views of the marshy eastern valley, with fog lying low between ridges. Spectacular! The trail then dropped me into Sages Ravine, while the incessant barrage of precipitation continued. Raging rapids in the ravine gave me pause, and I spent the next 30 minutes looking for the best way to get across. I finally settled on a plan, but the rocks along the route were rounded and slick with green slime, or partly submerged by the swift flow of the cold rushing water. The stream crossing was a very tricky, as I zigzagged my way over the assortment of boulders, without stepping in deep water.

A sign posted on the other side of the stream indicated I was now in Connecticut. From that point it was a steady climb up to Bear Mountain. I had repeatedly heard how easy the terrain would be in this state, however this proved to be a rather difficult climb, somewhat reminiscent of the Mahoosucs in southern Maine.

Nearing the higher elevations, there were a series of very steep exposures of rain-slickened Cambro-Ordovician quartzose-argillite boulders over much of the trail. A lot of slow cliff climbs were required up these rocky outcrops. Argillites are primarily composed of metamorphosed clay particles, which because of their platy texture, make this rock slicker than schist. To make it even more interesting, the perpetual drenching by fall rains helped the swift decomposition of the fallen leaves, making for an even more slickened surface over the exposed bedrock. Needless to say, I ended up on my backside many times on this trek.

There were rewarding views at the top. However, many of the overlooks were fogged in. My cell phone battery was at 12%, so I shut it down. This meant no photographs; I had to save battery life for an afternoon phone call to arrange for tonight's stay in Salisbury. Coming down Bear Mountain, it was an easy, pleasant hike all the way out to CT 41. The rain never stopped all day, and I was pretty much completely soaked through. I kept moving quickly along the trail, skipping lunch and any other breaks until about midafternoon, when I decided I really needed to stop and eat something. I tore open the Ziploc containing the tortillas, cheese sticks, Clif bars, and gorp that I packed for lunch. I ate as fast as possible standing next to my pack with the steady rain falling on me.

By the time I resumed my forward advance, I was cold again and just wanted to get off the mountain. Following a two-hour slip and slide fest on wet leaves and slick rocks, I reached the Route 41 parking lot. There was a large AT information board with a tiny bit of an overhang that I hunched under to get out of the rain. I called Vanessa, and in a few minutes, she cheerfully picked me up on the highway. When we arrived at her beautifully restored 300-year-old authentic colonial home, I unpacked my stuff, met her two friendly dogs, did laundry, showered, and visited the local market for my resupply. In addition, I bought lots of fruit and ice cream, and later, ordered a large cheesesteak from a nearby Italian restaurant. Mizza's, recommended by Vanessa, was in the next town over but was well worth it!

Later, I made a few phone calls to my family and turned in for the evening. I had the entire bunkroom to myself, which was exceedingly convenient so that I could organize and dry out all of my gear. There was snow and sub-freezing temps predicted for all night and through tomorrow. My feet and legs really needed a break, and I was considering taking a ZERO day right here.

Day 131
13.0 miles

Chapter Twenty-Three
Winter Arrives in Connecticut

Day 132 October 30, 2020 Friday
CT 41, Salisbury, CT to CT 41, Salisbury, CT
    Today was going to be a ZERO day at Vanessa Breton's house. In the morning following a quick breakfast, Vanessa kindly shuttled me to get new Merrell hiking boots and a Gore-Tex rain jacket at Barrington Outfitters, in Great Barrington, Massachusetts. Knowing exactly what I needed, I spoke briefly to an experienced trail hiker working there and was literally in and out of the store in less than ten minutes. When I returned, I needed to rest my sore feet and legs and the remainder of the day consisted mainly of R&R, ordering take out from Mizza's, and making phone calls home. Throughout the afternoon, the aroma of Vanessa's French cooking wafted through the house. For many years she owned a well-known French restaurant in town. Since its closing, she continued to daily prepare her delicious meals for private families nearby. Outside it was still snowing and 23°. Excellent choice to sit this one out!

Day 132
0 miles, ZERO day

Day 133 October 31, 2020 Saturday
CT 41, Salisbury, CT to Caesar Brook Campsite, CT
    My constellation alarm sounded off at 5 AM, I organized my pack, and then ate breakfast in the kitchen. I was ready to go by 6:30, waiting for Vanessa to come downstairs. She shuttled me to the AT trailhead where I had left off two days ago. It was a nice clear morning to hike but still quite cold at 25°. We said our goodbyes and I thanked her for the much-appreciated early ride. I headed south on the trail through the woods, past some neighborhoods, and through various open fields near Salisbury. It felt great to be out hiking again, especially with my new boots and Gore-Tex jacket. I finally felt like I was equipped for any kind of weather coming my way. The new Vibram soles had a noticeably sharper bite in every step, and I was anxious to test them on more Mahoosuc-style climbs later this week. Also, after more than 1,800 miles and three pairs of low cut Moabs, my erudite salesman from yesterday wisely recommended the higher-cut waterproof Moabs for these more demanding winter conditions.
    I made it to Housatonic River Falls by about 10 AM, mesmerized by the power of the torrent of water gushing over the impressive waterfalls and rapids at Great Falls. Prior to being deeply buried, compacted, folded, and metamorphosed into quartzite and marble, these rocks were

originally deposited along coastal proto-North America, about 450 million years ago, as carbonate muds and sandstones—similar to the sediment found today between Florida and the Bahamas. This was definitely not a river that I would want to navigate in my canoe! Delightfully, I met quite a few hikers along the path this morning as the trail weaved along the spirited river and near several small New England settlements. I passed by a busy food drive at a local high school, and several people stopped to ask me if I needed a gallon of milk, a dozen eggs, or bags full of canned goods. If weight of my pack were of no consequence, I would've heartily accepted the donations, but I politely declined all the kind offers. Soon, my lovely Housatonic Valley walk ended with a return to a forested uphill climb toward Belter's Bump. For the next ten miles or so, I was rambling through the snowy hills of the Sharon Mountains.

The upland trek thru these charming mountains was like a walk through a winter wonderland, with a soft white covering on every surface in sight. The warmth from the overhead sun was slowly melting yesterday's blanket, intermittently dropping clumps as I walked. I kept my hood up to prevent snowballs full of wet and cold from dropping down my neck and back. As I sauntered along, I could hear the mounting roar of small engines that seemed to be coming from somewhere in the valley below. When I emerged at a viewpoint near Easter Mountain, I could see down to a large racetrack where several small racecars were speeding around curves and straights on this sunny Saturday. This was Lime Rock Park, a popular racetrack where local race enthusiasts come to watch or participate in various styles of auto racing.

After an enjoyable day of touring these majestic woods, I settled in at a pleasing camp at Cesar Brook around 5 PM. I set up my tent, made dinner, and found a suitable bear box for my food bag. The water source was Cesar Brook, and a steady flow of cold, clear water made quick work of filling my water bottles. While inspecting my expansive campsite home for the evening, I followed the path to the privy to find nothing more than a small wooden box with a toilet seat on top. Still better than digging cat holes!

It was already becoming quite chilly, and the temperatures were promising to hit freezing by early morning. I brought both water bottles and filter into the tent, and stuffed my iPhone, charger, and Sawyer filter into my sleeping bag. Today had been a productive, long day, and I felt like it had made up for yesterday's ZERO day. Beginning tomorrow, there would be no more daylight savings time, so I would have to get up as early as 3:30 AM to put in the same number of daylight hours.

Day 133
19.1 miles

Day 134 November 1, 2020 Sunday
Caesar Brook Campsite, CT to Mt. Algo Lean-to, CT

I woke up at 4 AM to a very cold morning—about 30°. When I looked up at the sky, I couldn't see any stars. It was looking to be another overcast day but at least it wasn't raining—yet. I was on the trail by 6:18 AM after having my usual breakfast. Rising earlier from the change to EST really didn't bother me because my schedule was controlled entirely by the sun, not the clock. The trail started out rough and rocky like it almost always was, and my bright headlamp provided more than enough light to see my way through the predawn darkness. At the bottom of the first hill, it looked like I was going to have to ford a stream with my new hiking shoes. I sent back my Altras, thinking I wouldn't need a second pair of shoes any longer, and it also lightened my load. It took me about 20 minutes to pick a route across, carefully studying every option. After deciding on my best way forward, I carefully stepped from rock to rock holding my balance and soon I was on the other side. A few of the rocks were not completely above water and so it rushed over my feet a couple times. However, none soaked through my Gore-Tex boots. Clearly, it was money well spent to keep my feet dry.

From here, it was a leisurely 4.5-mile walk along the Housatonic River. With the ever-changing temperament of the river next to me, it was a wonderful hike with great scenery all along this section of trail. Because of the flat, smooth nature of my path this morning, I could snack often, with ease, while I hiked along—saving valuable daylight. I made it to Saint John Ledges Trail about 11 AM, and by then I had already devoured my entire lunch. There were lots of tourists, and day-hikers with plenty of inquiries about my journey. These were real nice folks, and I probably spent more time than usual engaging in unceasingly lively chitchat. The trail from the river road to the ledges was steep and rocky in places as I navigated amongst the exposed granite and metasediment escarpments and ledges.

Soon after, I decided to get to Mt. Algo Lean-to as soon as possible, as rain was forecast for mid-afternoon. It began drizzling around 2 o'clock, slowly intensifying. I made it to the shelter within the hour, just before the monsoons commenced. On my way up I noticed a pack alongside the trail, and to my right about a hundred yards or so, were two people under an enormous blown-down tree. They appeared to be noisily arguing about something, occasionally waving their arms. Beneath one of the massive branches, I could make out a man and a woman. The man was apparently the owner of the pack, and the woman was carrying what appeared to be a sack you'd expect to see slung over Opie's back when he feigned running away from home in Mayberry. My imagination ran wild with theories about the unfolding event, but the stronger driving force was the knowledge that if I didn't hurry, I would soon be caught up in the incipient and certain downpour.

Arriving at the shelter, I hurriedly ducked under cover, unpacked, and set up my air mattress and bag for the night. Occasions such as this made me truly appreciate the value of these shelters. As I boiled water for my noodles, I was captivated by the deluge around me. It was a comforting feeling knowing that my timing was right, and I was able to remain dry—at least for the time being!

No other visitors arrived night, although I kept expecting one, or both, of the strange couple from earlier to suddenly make their appearance, finally revealing the nature of their prior animated sideshow. My mind raced back to southern Virginia where reportedly, in 2019, a mentally unstable hiker brutally attacked a woman and murdered a man with a machete. Later, I slept alone in Mt. Algo Lean-to, listening to the amplified sound of every drop of rain, acorn, or branch that hit the tin roof, all the while listening for footsteps. It was a pleasant clatter that I could easily baffle with my earplugs, and I slept well. Tomorrow I was looking forward to better weather, with plans for reaching New York by noon.

Day 134
13.9 miles

Day 135 November 2, 2020 Monday
Mt. Algo Lean-to, CT to Wiley Shelter, NY

It was a 4 AM wake up, and I was thankful that last night's icy rain stopped sometime after midnight. Overhead, I looked up to see a near full moon and swiftly moving clouds approaching, suggesting more rain or snow on the way. Crisp black shadows of trees cast against shimmering rain-drenched earth surrounded me. By 5:18, I was back on the trail for yet another day, quickly shifting into a fast, body-warming, uphill pace toward the summit of Algo Mountain. As I approached the ridge top, filtered colors from the emerging sunrise were beginning to paint the sky bright pinks and orange, seemingly just for me to experience.

It was pleasant hiking all morning, offering frequent views of the Housatonic River over my left shoulder. My trail eventually led me down to river level around 10 AM near the town of Bull's Bridge. I followed a side trail for half a mile to Bull's Bridge Country Store, crossing an old, covered bridge on the way. When I arrived at the store, I could see it was very hiker friendly, with covered picnic tables outside, a spigot for filling water bottles, receptacles for charging cell phones, and even a hiker-amnesty box next to the front door. While there, I bought food to last through the weekend, plus lunch from the deli fridge, including two chilidogs and a wrap. Not exactly gourmet food, but to me it was like heaven. I returned to the trail and by 11 AM was once again strolling along next to the Housatonic River, gripped by the intensity of the class-five and -six rapids, inescapable in this feisty river. I won't be canoeing this section anytime soon, either.

The hike up and over Tenmile Hill was next, and then I'd finally be in New York. Up ahead was a wooden message board positioned for hikers entering Connecticut, with a scribbled inscription on the back indicating that I had just crossed the state line and was now in New York. At the next road, I came across a detour due to a bridge out over a supposedly non-crossable creek. I obediently followed it, walking along roads for several miles. Where the detour finally crossed that same creek, it was no more than three feet wide—I could have easily jumped across it in one leap!

After passing many homes and farms, the detour merged back with the trail. I hiked up a hill for about a half a mile to Wiley Shelter, and on the way, I spotted an old hand pump and used it to refill my bottles. Although it was from a well, I still filtered the water to be sure I didn't ingest any of the potential bacteria, protozoans or parasites, which could ruin your day. I considered tenting, but concerned about rain forecasted for the early morning, I decided to just stay in the shelter once again. I was in my bag by 6 PM. Tomorrow was Election Day, and it was on my mind. I ordered a mail-in ballot and it was sent to my home. Unfortunately, as I never knew where I'd be at any time along the trail, I never had it forwarded. Perhaps soon I would learn the results of the first presidential election I had not participated in since reaching voting age. Leave the politics at home. Glad I'm on the AT.

Day 135
12.4 miles

Chapter Twenty-Four
New York Highlands and Crossing the Hudson

Day 136 November 3, 2020 Tuesday
Wiley Shelter, NY to Stealth campsite on Hosner Mountain, NY

After a 3:30 AM wake up, I efficiently completed all morning chores, packed up, had a quick cold breakfast, and was ready to go. I was on trail by 5:15 AM but could probably do even better than this tomorrow. Terrain was gentle all day with a few boulder scrambles but the trail was mostly smooth and flat. Free from those pesky rocks and roots, I made great time. The key was to have lots of food and water ready, and to eat while I was moving. I took very few breaks, and none longer than five or ten minutes. My plan for tomorrow was to do a repeat with improved efficiency.

My morning trek took me through several expansive farm fields, with trails well marked by 4 x 4 posts and white blazes. One of the corncribs I passed appeared to have several windows with curtains and looked like somebody's home. Later, along the road north of Pawling, New York I came upon a railroad crossing. I turned my head to see a platform and waiting area for train passengers, either coming to, or leaving from, the Appalachian Trail. After crossing the tracks, the path led me to another boardwalk, crossing the wet and boggy, Swamp River Valley. Beavers once again came to mind. Nearing the end of the swamp, the boardwalk led to a bridge crossing over a small creek by the same name.

Later that morning I passed a picturesque lake by the unlikely name of Nuclear Lake. Apparently, an experimental nuclear research lab existed here at one time. The short version of the story I heard was that one day about 50 years ago, an accident occurred at the facility, causing some weapons-grade plutonium to be flung into the lake following a chemical explosion that blew out two windows in the laboratory. Following extensive cleanup and restoration, the lake and surrounding area was eventually determined to be clean and safe.

After lunch I crossed over I-84, and while on Stormville Mountain I came across a sign attached to a tree proclaiming, "Stormville Mountain Great Divide: From here north all water flows downhill;" comedians were everywhere. Shortly after this, I crossed over Hosner Mountain Road on the way up to Hosner Mountain. The trail followed the ridgeline and quickly brought me about 400 feet higher, where I found a very compact, but suitably flat campsite, with a lovely eastern view of the adjoining pastoral valley. It was precisely what I needed. All along the trail up to my campsite there were outcrops of amphibolite, a dark metamorphic rock containing little or no quartz, unlike gneiss. This type of rock results from the metamorphism of mafic igneous rocks like gabbro or basalt. Curiously, some of the adjacent rock units contain thick layers of limestone and dolomite, which are both sedimentary, suggesting later emplacement of these igneous rocks either as sills or lava flows.

After dinner I retired to my tent following the second-longest-mileage day of my journey. As I lie in my tent reviewing the day and planning for tomorrow, I could hear the far-off but relentless drone of traffic from I-84 and the Taconic Parkway in the valley below. I had only run into a handful of other hikers all day, but I crossed several highways and heard more traffic noise than any other day I could remember. I suspect this won't be the last time that I will make this proclamation, as I move deeper into the civilized world. Today was cold and crisp, but mostly clear and sunny equaling absolutely perfect hiking weather. If I could keep up this kind of pace, I could look forward to finishing in just two more weeks, and before Thanksgiving!

Day 136
23.1 miles

Day 137 November 4, 2020 Wednesday
Stealth campsite, NY on Hosner Mountain to Stealth campsite on Stone Walls Mountain, NY

I woke up at 3:30 AM and ate a cold breakfast and coffee in an attempt to conserve fuel. It was bitter cold and windy up here on my mountain, but I forced myself to move ahead. By 5 AM I was traipsing along the blazed trail in the dark, with my headlamp illuminating the narrow stony footpath. It led me along many open ledges of assorted Mesoproterozoic gneisses, overlooking the slumbering blackness of the valley below. Without a bright headlamp this treacherous path could be fatal. I could see and hear the distant wearisome line of commuters devotedly journeying toward their daily calling. This morning's pace was sluggish, and I was tired and sore from yesterday's 23-mile marathon. I wanted to repeat my near record again today but was just not feeling it.

I met several hikers on trail this morning, including three New York gentlemen about my age who were out for three to four days. We stopped and talked for several minutes, exchanging information about the trail ahead for each of us. Later, I chatted with a young man, and told him how I was feeling unusually fatigued today. After listening, he explained that he could make his energy transfer to me, thus making me feel more invigorated and energized. Kind of like AirDropping some Geritol. I thanked him for his sincerely benevolent gesture.

At about 8 AM, I arrived at RPH Shelter, or Ralph's Peak Hikers Cabin. It reminded me more of a college dorm room than an AT shelter, but was unique, clean, and tidy. Inside was a supply of several gallons of spring water, dozens of paperback books, and chairs—and even a desk.

Outside was perhaps the nicest privy I'd seen on the entire AT. Following this, the initial climb to Shenandoah Mountain over the gneiss and quartzite boulders was steep and difficult, but once I made it to the ridgeline, it was just a steady but rocky climb for the next hour or so.

Throughout the morning I continued consuming an assortment of energy bars and drinking copious amounts of water, coffee, and caffeinated drink mixes. My path took me past numerous glacial ponds and marshes, eventually leading me along a particularly pleasing part of the trail. For the next several miles, I had nearly continuous views of the lovely Canopus Lake. I crested one of the grassy, sunlit knolls around 2 PM, where I stopped for a short rest against a perfectly positioned gneiss boulder, feeling once again like I was back in summer, if only for a brief interlude. After about two more miles, I fortuned upon a pristine campsite nestled between two distant escarped ridges. The bedrock was billion-year-old sub-metamorphosed amphibolite, a type of rock found typically in oceanic crust. For the past several miles I had been rambling past centuries old stone walls constructed from these very same rocks. At one time, this place must have been cleared as pastureland or grain fields by early settlers, but since has returned to a dense forest of mature hardwoods.

I set up camp around 4:30 and was soon getting ready for dinner. Tonight's was ramen surf and turf, exquisitely made with the addition of tuna and chunks of beef jerky mixed in with my regular noodles. Also included were some cranberries and jalapeño-cheddar crackers, for enhanced texture and flavor. It was actually quite delicious! While I was preparing this well-deserved feast, a playful yellow-lab puppy came running up and licked my face and arms. I quickly picked up my culinary masterpiece and held it firmly in one hand so my excitable new furry friend wouldn't knock it over. He belonged to an affable young lady out for her afternoon walk. She briefly came over for a little chat while re-establishing calm in her canine companion.

By the time I had finished my supper and hung my food bag, the sun had sunk below the horizon, and the sky was a golden-orange glow. By 6:30, I was in my tent for the night. Upon inspection, I found three deer ticks on my skin from earlier today, likely when I sat down to rest on the knoll. It had been a while since I removed one or two of these stealthy critters following a day in the woods, but it had also been a while since it was this summerlike. It had been a very long day—much more strenuous than yesterday. I calculated that I would need to average about 20 miles a day in order to finish by mid-November. I was definitely up for the challenge.

Tomorrow morning, I would be stopping at the Appalachian Deli on U.S. Highway 9, only four miles away. A 5 AM departure should get me there at 7 AM; really something to look forward to!

Day 137
17.9 miles

Day 138 November 5, 2020 Thursday
Stealth campsite on Stone Walls Mountain, NY to William Brien Memorial Shelter, NY

I woke up at 3:40, had cold coffee and bars for breakfast. My right knee was bothering me again. I hadn't worn my elastic brace for about a week as it was difficult to pull it over my Merino woollies. For about an hour into my hike through a particularly rocky section, persistent pain was bothering me with each stride. Eventually, I stopped to shed the woollies and put on the knee brace. It helped, and although not a 100% improvement, it was noticeable and from here on out I would continue wearing it.

After several more miles, I emerged from the forest, temporarily stepping back into civilization, to see and hear morning traffic on Highway 9. Across the street was my long-anticipated Appalachian Deli. I strolled over and entered to find a dazzling menu of every possible breakfast delight imaginable, with names like: "Big Mess," "Heart Attack," "Diablo," and "Burrito Overload." For the next 20 minutes, I heartily enjoyed my heavenly conglomeration of eggs, hash browns, fried peppers and onions, corned-beef hash and hot coffee. After this magical gluttonfest, I went to the outside spigot to fill my water bottles. When I took a swig, I immediately spit it out. It tasted like chemicals and old pipes. I went back into the deli, asked the manager about it, and she scurried back into the kitchen with my bottles. She returned with the two both cleaned and filled and apologized explaining the Deli's system was just re-plumbed. Apparently, the outside spigot had not yet been flushed. I thanked her and soon was back on the pathway to more adventure and excitement ahead.

Around 10 AM after a long, winding descent from the New York highlands, I arrived at the mighty Hudson River. Ahead of me and across the busy highway was the famous Bear Mountain Bridge, beyond which its namesake stood towering at a modest 1,289 feet above sea level. In less than one-half mile, I would soon be standing at the official lowest point on the entire AT, a mere 124 feet above sea level. As I approached the nearly half-mile-long suspension bridge, I caught majestic glimpses of the Hudson Valley looking north and south with the backdrop of autumn foliage blanketing the surrounding hills. Crossing the Hudson I could see age-old railroad infrastructure still in use today, belying the age of some of the tunnels and trestles built for this early mode of transportation.

On the other side of the river was Bear Mountain, a popular tourist spot surrounded by a well-manicured park-like setting, including the beautiful Hessian Lake in its foreground. The mountain is underlain by massive gneiss and granite of the Neoproterozoic-age Storm King Granite, which forms a durable pavement at the summit. I continued following the trusty white blazes, eventually leading me to the start of the popular ascent to the pinnacle of the main attraction.

The hike to the top was fun, consisting of over 1,000 granite steps to the summit of Bear Mountain. In full 40-pound pack and with now nearly five months of daily training behind me, I was able to complete the entire climb without once stopping, passing every other uphill trekker I encountered. It was comparative challenges such as this that made me aware of what great shape my two legs were in. The view from the top was, of course, incredible. I gazed south toward the barely visible skyscrapers of the New York City skyline in the faint horizon. I spent about 30 minutes at the summit engaging in polite dialogue with other tourists and day-hikers. I met only one other thru-hiker, by the name of Hopscotch, who just finished a flip-flop two weeks ago and was back doing some local hikes for a couple of days.

For the remainder of the day, I trampled across more scenic mountaintops with numerous vistas directing my attention back to the scenic Bear Mountain and its neighboring peaks. At 5 PM, I arrived at the William Brien Memorial Shelter, amidst a craggy ensemble of billion-year-old metamorphics piercing their way skyward out of the Earth. The rugged edifice was built out of the very same rocks and looked, to me, more like an Armageddon-worthy bunker than an AT shelter. By now, the sun was low in the sky, and I expeditiously found a nice flat tent site nearby and made camp for the evening. I sat on a large fallen-oak log to boil water for my noodles in the still warmth of the slowly setting sun.

These much-appreciated unseasonably warm days were predicted to last another week, perhaps long enough to allow me to finish my voyage before having to wake up one morning covered in a thick blanket of snow.

Day 138
19.9 miles

Chapter Twenty-Five
Boardwalks, Stairways to Heaven and Water Gaps

Day 139 November 6, 2020 Friday
William Brien Memorial Shelter, NY to Lakes Road (pickup for motel), NY

I awoke 3:30 AM to one of the warmest days in a long time. The morning weather certainly did not feel like November. My plan today was to meet some of my Pittsburgh geologist friends, Steve, Robert, and Bob later this afternoon for a weekend visit. So, I worked hard in the morning to get in as many miles as possible. The morning hike included lengthy exposures of glacially smoothed gneiss bedrock, generously exposed along ridgelines. This was very much like walking on sidewalks. Clear skies and sparse foliage allowed for countless picturesque vistas along these mountaintop rambles.

Later, there were several difficult down climbs where deep crevasses filled with giant boulders required slow and careful maneuvering. Many of these steep-cliff paths were now covered in leaves and fallen trees, adding to the difficulty of the descent. One of the better-known challenges was the Lemon Squeezer, a collage of metavolcanic boulders, cliffs, and tight caves offering a formidable afternoon obstacle course following my abbreviated lunch break. Slick, decaying leaves carpeting the hodgepodge of car- and truck-sized boulders made it hard to move quickly. Most of the hill climbs and descents were on the order of 150 to 300 feet and not overly severe, but they were plentiful and wore me out. Additionally, there seemed to be a multitude of poorly marked trails where past hikers had been unsure of the correct route, often resulting in worn trails taking off in several directions.

Around 4 PM, I came down off the last ridge top trail to Fitzgerald Falls, an impressive cascade over moss-covered Proterozoic gneiss ledges into a clear pool at its base. When I arrived on Lake Road, the hometown gang was already there and waiting for me. We exchanged the usual wise cracks and greetings and headed back to town. Later, I was treated to a scrumptious dinner (which incidentally did not include any ramen or pre-packed tuna fish) at the Sit n' Chat Diner in Sussex, New Jersey.

Day 139
18.5 miles

Day 140 November 7, 2020 Saturday
Lake Road (drop-off from motel), NY to Long House Road (pickup for motel), NJ

I woke up at Rolling Hills Motel at 6 AM, having enjoyed the luxury of sleeping in a real bed the night before. All four of us got up and headed out for breakfast at 7 AM, returning to the Sit n' Chat Diner. Our favorite waitress was already there and cheerfully took our orders. We all discussed our plans for the day, including where to meet after the day's hike. After demolishing our morning gluttonfest, Bob and I arrived at the trailhead at 9:30, with plans to slack pack along a New York ridgeline that would lead us into New Jersey by midafternoon. This would be the second time this year backpacking with Bob, the first ending in August at Boiling Springs, Pennsylvania. Since our first adventure, he continued section-hiking across the Cumberland Valley to points eastward, on and off again, throughout the next three months. Bob eventually finished all of Pennsylvania by mid-November with plans to continue as a NOBO to Katahdin in 2021, possibly with his son Nate.

It was a great day for being on the trail, sunny and 75° by midafternoon with spectacular views all day long. Because it was a weekend, we encountered many tourists and day-hikers along the way. Our climb to the top of Bellvale Mountain from Lake Road was expeditious, and our morning hike was mostly along a glacially smoothed bedrock ridgeline comprised of steeply dipping to overturned Devonian Bellvale Sandstone. In the valley to the east, Oriskany Sandstone, my trail namesake, was present in outcrop paralleling Lake Road and then trending beneath Greenwood Lake. I was out of the metamorphics and back into the Appalachian sedimentary basin again, but only for today.

After lunch, Bob and I continued south along the apex trail, stopped for water at a campsite, then continued south toward the New Jersey state line. I was about a quarter mile ahead of Bob when I noticed a young lady sitting on a sunny rock with a notepad. I stopped to say hello and she began explaining that she was looking for an idea to make a YouTube video. I said, "How about doing something about the rocks, since they are all around you?" She responded, "I suppose I could do that since I'm a geologist." I exclaimed, "Me too!" Alex, a young geologist originally educated in Poland, came to the US several years ago to complete work for her master's degree and now works in Connecticut. Moments later, Bob joined us and the three of us rockhounds chatted for the next twenty minutes about the geology beneath our feet.

Toward mid-afternoon, Bob and I crossed over the state line to New Jersey and then met up with Steve and Robert around 4:30 at Long House Road. Later that evening, we all had a delightful dinner at the Brick and Brew Restaurant in Franklin, New Jersey.

Day 140
11.8 miles

Day 141 November 8, 2020 Sunday
Long House Road (drop-off from motel), NJ to Pochuck Mountain Shelter, NJ

My alarm woke me at 6 AM in the motel, and we all drove back to the Sit n' Chat Diner, arriving before 7 AM before the doors were open. We waited a few minutes and then were the first in, once again consuming mass quantities of food from our favorite restaurant—this time breakfast goodies. After the feasting, we said our goodbyes and the crew dropped me off at the trailhead on Longhouse Road around 9:45. Today was once again warm and sunny with a high predicted for 78°, very unseasonable for November, but I was ecstatic to have these nice days to enjoy on the trail.

I began trekking towards Pinwheel Vista on Wawayanda Mountain, where the exposed bedrock was primarily massive, foliated pyroxene granite and gneiss. I was back in the metamorphics for a while, but near the thrust front, similar physiographically to the Blue Ridge Front of Virginia. As I approached the summit, I noticed the number of tourists suddenly increasing. When I finally reached the main vista, there were dozens of tourists milling all around as I continued without a stop toward the downward exit trail. Coming down the mountain, the trail was essentially large granite steps known as "Stairway to Heaven." Robert Plant echoed in my head as I considered this rock-on-rock irony. There were literally hundreds of tourists of all ages, shapes and sizes going up and down at the same time, making this the absolute single busiest day on the AT since the start of my quest—nothing before even came close to the crowds I experienced today.

After I made my descent into the Cambrian dolomite of Vernon Valley, I soon entered a marsh area where wet, boggy conditions required the construction of a four-foot-wide boardwalk for the next several miles. Glacial origin, later enhanced by the beaver engineers, I surmised. After crossing the wide wetlands valley, I eventually climbed Pochuck Mountain, where I made it to Pochuck Shelter by 4:30. After having been part of the largest crowd I encountered all summer, it seemed odd that I was alone here at this shelter. I refilled my water bottles, set up camp, made dinner, and was in my tent by 6:15.

Day 141
13.8 miles

Day 142 November 9, 2020 Monday
Pochuck Mountain Shelter, NJ to Mashipacong Shelter, NJ

Forgetting to set my alarm, I woke up at 5:30 AM, resulting in a late start. I packed up, had quick breakfast and was on trail by 6:30. After an initial half-mile downhill, my trail was essentially flat for the next several miles. As I strolled along the marshes and over the many boardwalks, there were hundreds of geese, ducks, and other water birds all around me. This was truly a bird-hunter's paradise, but I wasn't here for that. It was fast and easy hiking, and I was making pretty good time. This wide valley was physiographically equivalent to the Lehigh Valley, where I was born and raised. As I continued over the long expanses of fields and low hills, elevations began slowly increasing toward High Point. I could see the monument on the far ridge from about eight miles away. At an elevation of 1,803 feet, this impressive stone structure was built on the highest point in New Jersey to honor all past war veterans and is constructed as a simple obelisk. While hiking to the top of the ridge where the monument is located, I encountered dozens of tourists and day-hikers. I was in a hurry however and did not stop at the observation deck.

Later in afternoon, there were some nice Tuscarora Sandstone ridge-walks with lots of great views of western New Jersey valleys along Kittatinny Mountain. I arrived at Mashipacong Shelter by 4:30 PM, set up tent, made Beef Chili Mac for dinner, and was in my tent by 6:30. All day it seemed that I was tripping over rocks and roots, as they were hiding beneath a dense covering of leaves. I somehow hurt my left calf muscle today and was desperately hoping it would be better by morning.

Day 142
19.6 miles

Day 143 **November 10, 2020 Tuesday**
Mashipacong Shelter, NJ to Stealth campsite on mountain, NJ

At precisely 3:30 I awoke to my alarm, made breakfast of hot oatmeal and coffee, packed up, and hit the trail by 5 AM. Surprisingly (and thankfully) my left calf felt fine this morning. Later, It began hurting for a little while after lunch break but soon got better. Was I not drinking enough water and was I dehydrated?

This morning's sunrise was glorious. There were open views to the valley beneath me from rock ledges and little elevation change as I moved along. I stopped at Sunrise Mountain just as the sun was coming up for an incredible lightshow of pastel pinks and oranges. I was in a rush to log in as many miles as possible today so that tomorrow would be easier, considering there was rain in the forecast. It became tedious walking on some of the exposed expanses of Shawangunk Sandstone. Many of the massive rocks were glistening with oozing moisture and were already slick with lichens. While hiking the ridges, I was constantly snacking on my lunch items, finishing all my food before noon.

That afternoon, I was worried the ridge walk would be brutal like so many other places where the trail followed Tuscarora, a near stratigraphic equivalent to the Shawangunk. I remembered struggling along Garden Mountain near Burke's Garden, or the knife-edge in Pennsylvania—all the same rocks! I was pleased to discover that most of the afternoon hike—about 10 miles—was on old gravel or dirt forest roads. This was indeed a nice surprise. I hiked until sunset, when I found a nice flat campsite on a breezy summit near the trail. All afternoon I was so famished that, when I finally established my night's camp, I gobbled down two freeze-dried meals: Sweet Pork and Rice, and Teriyaki Chicken and Rice. Later I tried to call Deer Head Inn for tomorrow's hostel stay, but there were no human operators answering the phone, only recorded messages. "COVID?" I wondered. I retired to my tent by 7:45.

Day 143
22.4 miles

Day 144 November 11 Wednesday
Stealth campsite on mountain, NJ to PA 611, Delaware Water Gap, PA

At 3:30 AM, I woke up to wind and warm fog on the mountaintop. There was no rain yet, but I could feel it coming. I packed up, guzzled my coffee, and was on trail by 5:15 AM. My route was very rocky this morning, and the fog made it difficult to see. Everything around me seemed mysterious and otherworldly. I passed several ponds. Some ponds were perched high on Kittatinny Mountain and some I could see off in the valley below. These ponds were remnants from deglaciation some 18,000 years ago when continental glaciers gouged out depressions on the landscape, or later left massive ice chunks to melt as the glaciers retreated north.

Today, I had my final destination of Pennsylvania in my sights. Time seemed to stand still. While plodding along atop Kittatinny Mountain I met a very interesting young lady running the trail. As she approached, we said hello, and she stopped. Chatting a few minutes, she said her trail name was Quick with a K, and further explained that she holds the fastest-known-time (FKT) record for New Jersey trail running on the AT. Her time for this amazing 144 mile out-and-back was 3 days, 19 hours and 52 minutes. Wow! Learning I was a thru-hiker, she got excited saying she was giving a talk that evening and asked if it would be okay to photograph me and mention my name with a brief bio. Of course, I agreed and was honored to be in the presence of such a positive-minded trail champion.

Later, I met three other trail runners around Sunfish Pond where the trail was insanely rocky with sharp, jagged quartzite boulders, also of glacial origin. These rocks were extremely difficult to navigate, and as I carefully hopped from one large boulder to the next, I wondered how long it would continue. If too long, would I have enough daylight to make it to Pennsylvania before dark? Eventually the trail improved, and my pace increased once again. After that pond, it was mostly smooth trail on either old forest-service roads or less rocky trails. Finally, the path descended to the Delaware River floodplain at I-80.

I took the I-80 walkway across the Delaware River next to heavy traffic buzzing past me within arm's length the entire way across the bridge. The trail led me to the small town of Delaware Water Gap and directly to the Deer Head Inn. I walked up the front steps to what appeared to be the entrance. I then checked all the doors, and they were all locked. I could see it was dark inside, as if no one was there, and the lights were out. So, continuing down the street, I walked a half-mile more to the Clarion Inn and checked in. I did my re-supply at minimarts across the street, dried out my sleeping bag in the dryer, and did my laundry without soap. The reason for this was that I spent all my spare change on the wash cycle and was a quarter short to purchase soap. Later, I ate dinner at Joe Bosco's barbecue joint next door, and the brisket was outstanding. I came back to my hotel, made a few phone calls, and inhaled a pint of Ben & Jerry's Cherry Garcia ice cream. I was in bed by 9 PM with aspirations to do at least 20 miles tomorrow.

Day 144
14.8 miles

Chapter Twenty-Six
Sun Up to Sun Down through the Keystone State

Day 145 November 12, 2020 Thursday
PA 611, Delaware Water Gap, PA to Leroy A. Smith Shelter, PA
Another 3:30 AM wake up, this time at the Clarion Inn in Delaware Water Gap. I left the hotel after a quick vending-machine breakfast consisting of a bear claw, Tastykake Krimpet, and coffee.

I was on the trail before 5 AM in the dark with my headlamp. It was 61° and there was a light drizzle. I walked through town toward the white-blazed telephone poles and street signs, and then followed them up a dead-end road to the trailhead where I began a steady upward climb along the hillside. The light from my Petzl illuminated a narrow patch of gravelly trail as I slowly gained elevation above the valley floor. The trail was tough to follow at times as heavy tourist traffic through the years created false leads everywhere. There were several breathtaking vistas on the way up, for example the view from Lookout Rock onto the river some 500 feet below, or from Mt. Minsi, at 1,100 feet above the rapids. At the summit there were many more impressive views along the Delaware River, with hundreds of headlights speckled amongst the slow-moving snake slithering along the river.

Once I reached the ridgeline trail, the terrain from thereafter was kind to my feet. However, the wind was growing stronger, the temperature was falling, and soon came the pattering of raindrops. I added a few layers to stay warm and dry and officially proclaimed an end to wearing shorts for this year! A while later, Wolf Rocks proved challenging in the heavy rain, as I tried to keep from sliding off the slick surfaces of steeply dipping orthoquartzites. There were a few more rocky segments, but generally these were not a big deal for me to negotiate. Occasionally, I could catch a clear glimpse through the trees to the southeast, toward the sleepy hamlets of Bangor, Roseto and Pen Argyl.

Despite the foul weather, I achieved my goal of 20 miles for the day, arriving at my shelter at 4:30. I swept out the inside, set up my air mattress and bag, and got water from a spring. I next boiled water for my noodles and was toasty warm in my sleeping bag by 7 PM. I was only 91 miles from finishing this long, storied adventure—I could almost taste it.

Day 145
20.3 miles

Day 146 November 13, 2020 Friday
Leroy A. Smith Shelter, PA to stealth campsite
(near Lehigh Furnace Gap), PA

A 3:30 AM wake up in the shelter gave me time to pack up, eat a hot breakfast, and be on trail by 5:15. The morning weather was foggy, drizzling, cold, and damp. The trail was generally easy, and I moved along quickly most of the morning. Near Little Gap, I passed Blue Mountain Ski Area, but it was totally fogged in with no views.

Later near Palmerton, I hiked through newly planted pines and grasses at an EPA Superfund site on Blue Mountain east of Lehigh River. From the vista points, I could see the same conifer plantings established along the once-bald ridge line on the west side too. In addition, there were many views looking into the valley where I could see the old New Jersey Zinc plant, a century-old factory responsible for zinc-sulfide emissions that killed the vegetation along neighboring ridgelines many decades ago. The sight of the old factory reminded me of the ecological toll of historical mining and ore smelting. These ores ultimately yield necessary metals and critical elements that all of us in the modern world take so much for granted. Ironically, people today desire the luxuries of modern technology, manufacturing and so-called "green energy" but abhor the idea of its extraction and processing in their back yard.

It had been raining much of the morning and my feet and legs were feeling weary. I decided to take the blue-blaze (bad weather) trail to the bottom instead of the regular boulder-scramble route, which I had done in the past as a boy. This was indeed an interesting mountain. The once moonscape of the summit had been transformed into a newly planted conifer forest. Also, there was a network of abandoned postindustrial infrastructure left in place—mainly roads—built during a remediation phase some 25 years ago. As my path led me down the mountain, I passed a great deal of fencing and netting installed to prevent rock falls onto the highway.

While crossing the PA 873 Bridge over the Lehigh River, I passed a sign notifying me I was now entering Lehigh County. This was a monumental milepost for me since I had now just crossed over into the county of my birth and upbringing. From this point forward, every step to the finish at Swatara Gap would be a repeat of trail that I once, twice, or a dozen times hiked in my past, trampled on during bygone adventures in my youth. Having crossed the Lehigh, I next followed the blazes across the highway to a trailhead leading me up Blue Mountain west of Lehigh Gap. On my way to the top, I met two ladies who were out for a day-hike, and we talked about my journey. They both said that someday they also wanted to thru-hike the AT, but for now they had far too many work and family obligations preventing them from such a trek.

The trail to the summit was gently graded, and when I reached the ridgeline, the nearly level corridor paraded me by an abundance of inviting campsites in this pristine pine forest. I rambled along until about 4:30 when I eventually stumbled onto my stopping point for the day, about one mile east of Lehigh Furnace. I was planning to meet my brother David and Mom tomorrow at 10:30 AM at Route 309, just nine miles from here. They each live about 30 minutes away from our rendezvous point and were bringing me a final resupply and some lunch. A 5 AM start tomorrow morning should be early enough to get me there on time.

Day 146
20.1 miles

Day 147 November 14, 2020 Saturday
Stealth campsite (near Lehigh Furnace Gap), PA to stealth campsite (near Dan's Pulpit), PA

Once more, my 3:30 AM constellation alarm cried out, waking me up for yet another day. I made my standard breakfast, packed up and was soon on trail by 5 AM. The morning was cool and clear as I started out in the dark with my trusty headlamp leading me steadily downhill to Lehigh Furnace Gap. Soon, the footpath crossed Blue Mountain Road, where some wonderful trail angels left several liters of fresh clean water near the AT sign by the road. This was much appreciated, since up on these ridges water was very hard to come by.

As I was trekking towards Bake Oven Knob, I noticed the path to the shelter on my left. As often as I had been to Bake Oven, I never visited this structure, so I thought I would check it out. It was an average AT shelter, and seeing all the gear lying around, I assumed the hikers had climbed up the rocks to watch the sunrise. Returning to the trail, I continued up towards the summit. It became increasingly rockier the higher I climbed, and I had forgotten what a prominent point this was along the ridgeline. I used to come here quite often with friends and family, both on overnight hikes and afternoon excursions, just for the impressive views of the valley. As I approached the knob, I could see several groups of hikers perched at various optimum-viewing points on the knob. When I completed my scramble up to my chosen perch, the sun had already come up over the horizon; I missed the fire-orange pre-dawn sky. It was still a glorious sight as well as a stunning view of the Lehigh Valley from this vantage point. All the boulders here were Silurian Tuscarora sandstone left in this pile by the plucking action of advancing and retreating ice sheets during the past 2.6 million years. I remained there for about 10 minutes, just peering off into the distance, looking for specks of familiarity in the panorama before me. I soon returned to the trail where I ran into several day-hikers coming to spend the morning watching hawks and other birds. One of them said to me that this was as good as Hawk Mountain for watching hawks, and the price was right too: free.

At about 10 AM I arrived at Route 309, a bit early for my rendezvous with Dave and Mom. I was anxious to see both of them and enjoy some good food. I sat down by one of the signboards, leaning against my pack for a short rest. After several minutes, I still didn't see anybody, and so I texted Dave. It turned out they were waiting for me in a parking lot across the highway. I walked across the road and found them talking with a few other hikers who just came off the trail for some water. It was good to see Mom and Dave, and they brought three ham sandwiches, orange juice, apples, soup, and radishes. We sat in Dave's car and talked while I feasted. After about an hour we said our goodbyes and I was back on the trail.

The next shelter I passed was named for the Allentown Hiking Club, which I also had never seen; I decided that I would check it out too. A few folks were there when I arrived, we chatted a while, and then I was back on the trail. I heard the incessant echoing of gunshots all day long. It was a Saturday and bird- and small-game season was in. Hunters were out six days a week shooting rabbit, squirrel, grouse, pheasant, and quail very close to where I was hiking. For this reason, I generally wore a bright-yellow pack cover, and a bright-orange Gore-Tex rain jacket. Archery season in Pennsylvania began before I arrived as well, so there were also lots of bow hunters out for deer. I rarely saw any of these well-camouflaged archers other than in parking lots, or along the trail towards the end of the day. Fortunately, opening day for regular firearms hunting for deer was not until the Saturday after Thanksgiving, so I didn't have to worry about that.

My goal was to get to the Eckville Shelter by sunset. Considering the time and the distance left to travel, I knew it wasn't likely. I used to always believe Eckville was named after someone from the Eck family—the namesake of my paternal grandmother's family. This knowledge had always made me feel at home here. However, the true origin of the name came from the location of this small hamlet's proximity to the corner of Albany township, and the German word *die Ecke*, meaning corner. Even knowing this, I still felt at home. After recognizing that Eckville was too ambitious a target, I set a new goal for Dan's Pulpit, a summit along the trail up ahead, hoping that there was a nice stealth site somewhere just past it.

As the sun was setting and I went beyond the summit, I found a perfect camping spot. It was flat and spacious, and as I peered off in the direction of Hawk Mountain, I could see the fiery colors of the setting sun through the sparse, leafless branches of the overhead trees. There was a quiet peacefulness surrounding me as I set up my tent and found the perfect tree to hang my food bag. In the remaining moments of twilight, I started my gas stove, boiled some water, made dinner, and was in my tent by 6:30 PM.

Day 147
18.6 miles

Day 148 November 15, 2020 Sunday
Stealth campsite (near Dan's Pulpit), PA to stealth campsite (near Auburn Lookout), PA

The 3:30 AM wake up from my dependable alarm launched me out into the cold to begin my third-to-last morning of camp routines. After breakfast, I packed up and was on trail by 5:05 AM. It was a nice hike down the mountain in the dark with my newly recharged Petzl. I bought this headlamp back in New Hampshire because both it and my iPhone used the same charger, a Tzumi Pocket Juice portable charger. On my way into the valley, I passed over several streams, which I assumed were coming out of the River of Rocks, a periglacial feature extending up the crease in the mountainside toward Hawk Mountain. I filled up one of my water bottles at a stream and continued on my way. Crossing Eckville Road, I noticed a sign indicating Eckville Shelter was a short distance down the road, and I considered taking a look but decided to just keep moving forward.

From here, my trek to the Pinnacle was mostly on old forest-service roads. As I was nearing the summit, I sniffed the air and instantly recognized the comforting aroma of a campfire. Soon I encountered two hikers camped out from the evening before, having their morning coffee around the smoldering collection of snapped twigs and branches. I talked with them for a while and learned one of the fellows had been a section-hiker for many years on the AT. We talked about places we'd both seen and even shared our near-death experiences from the White Mountains. Of all the thru-hikers I've talked with, many have at least one bizarre story to tell about the Whites.

I left them around 7:30 AM, and on my way to the overlook I passed several more hikers, several mountain bikers, and four horseback riders. At the summit, I chatted with lots of folks including a lady and her spirited young son from Philadelphia, who were just leaving to get back to their car before the next deluge. On the way down the mountain, we would pass each other several times. It was kind of funny and the young boy was entertained by our repeated encounters, laughing harder each time.

Roughly a mile south on the AT past the Pinnacle, there was another overlook known as Pulpit Rock. Just behind it is Pulpit Rock Astronomical Park. Originally built here in the 1960s, this park was located here in order to be far away from the light pollution of the valley. It was later renovated, and new buildings were added in the 1990s. The entire astronomical campus went largely unnoticed, at least to me, in the frequent jaunts here in my earlier years. I honestly don't ever recall seeing it. As for the Pinnacle, I would often hike to this high point and stunning overlook with friends, my church group, or my high-school natural-history club. What is so special about this particular vantage point, geologically, is its position with regards to bends in the structural grain of the fold-and-thrust belt, better known as the Ridge and Valley Province. The primary ridge-forming unit holding up the entire ridgeline along Blue and Kittatinny Mountains is lower Silurian Tuscarora, or toward New Jersey, Shawangunk Sandstone. These rocks include very pure quartz sandstones whose grains are often cemented together by silica cement creating some of the most resistant-to-weathering rock formations in the entire sedimentary section of the Appalachian Basin. These especially quartz-rich sandstones are commonly known as orthoquartzites, and account for why hikers often groan when considering hiking the A.T through Pennsylvania. In addition, there are remnant periglacial deposits scattered all throughout the Keystone State, some covering prominent ridgeline areas where seasonal freeze-thaw cycles during the last ice ages broke up these massive, tough orthoquartzites and either left them in their original position, or scattered them wildly as gravity, precipitation, and slope angle of the ridges dictated.

Standing here again felt very familiar, and although I wanted to hang around and take in the scenery, I knew that I had to keep moving. Perhaps my most memorable trip to the Pinnacle was with my friend Ted, when we were in High School. Ted and I planned a couple of days backpacking on the AT, and my mom had recommended premixing pancake batter and storing it in Tupperware for the next morning. We began our hike around sunset not considering the difficulties for which we were setting ourselves up. We started up a trail from the southeast bringing us right up the steepest part of Blue Mountain, which by this time, we were climbing in the dark. When we finally arrived at our campsite, Ted began unpacking only to discover pancake batter coating the entire contents of his backpack!

I wanted to try to find a campsite well past Port Clinton before the rain hit. I was pushing hard to keep good pace all morning. The hike down to Schuylkill River Gap was along a moderate grade and fairly easy. When I reached the quiet hamlet of Port Clinton it seemed like a ghost town, no one was out as I sauntered through the side streets following the white blazes across the Little and (Big) Schuylkill Rivers. I crossed over some railroad tracks, spotting the sign for the AT as it climbed a convoy of rock steps straight up the mountain. When I looked down at one of the first steps, I was surprised to see my good friend Arthrophycus, which I hadn't noticed since that cloudy day up on Garden Ridge in Virginia. It was after 1 PM and the ascent up the mountain from Schuylkill Gap was about as steep as any I scaled in a while. It was mostly rock steps, so although many of the emplacements were coming loose as I ratcheted my weight upward, time passed quickly, and before long, I was exultant at the top.

I continued trudging along the monotonous footpath as it laced me through the ridgetop timberland. It was a dreary day with no signs of life left in this frigid autumn forest. After about a mile of rambling on this flat, I spotted a proper stealth site. The winds were picking up intensity, and the rain was just beginning. I was able to get my tent set up and gear inside without getting anything too wet. I quickly finished dinner and hung my food bag in a very light drizzle before finally crawling into my tent. Soon, the ridiculously insane monsoon rains and typhoon winds suddenly detonated for the next six straight hours. This unbridled barrage of precipitation and accompanying ear-splitting percussion-laced lightshow continued throughout most of the night. I was once again eternally grateful that my timing had worked out, and that I was ensconced in the miracle of my Dyneema shelter.

Day 148
18.9 miles
Only 33 miles to go!!

Day 149 November 16, 2020 Monday
Stealth campsite (near Auburn Lookout), PA to
Pilger Run Spring Trail (Appleby Campsite), PA
    I woke up to my trusty alarm at 3:45 AM. Skipping a hot breakfast, I was on trail by 5:10. Weather was cold (40°) but clear. Stars were out and the rain was gone as I hiked several miles in the dark while watching a beautiful sunrise. By about 8:30, I intersected some nice mountain streams and decided to stop for some water and make that hot oatmeal and coffee I missed this morning. After my morning break, I came across several stretches of boulders where I hopped from one rock to the next. Fortunately, my pack had become quite light, as very little water and food remained. After I crossed Route 183, I stopped for a quick lunch, and then started down the trail where I soon encountered two elderly section-hikers. We talked for a couple of minutes, exchanging information about the trails ahead of us.
    Later, I arrived at an old campsite where we used to swim in an ice-cold spring-fed pond just down the trail a ways. I camped here maybe half a dozen times in the past and it brought back a lot of good memories. There was a young man camping alone here, and we talked about some notable sections of the trail we'd been to recently. I told him not to miss the pond experience. The remainder of the day was very pleasant—cool and sunny. I thought a lot about how I would miss this daily routine, but I was ready to finish. As I continued south, I came across many great viewpoints from the overlooks along the way. This was my last long marathon day and I tried to take in all the sights and sounds along the way. Soon, I would be getting up every morning in my own bed and having a cup of coffee on the leather couch with my wife and our two cats in a warm, dry house.
    I arrived at Applebee Campsite right at sunset, set up my Altaplex tent for the last time on the AT in 2020. Of course, I had ramen noodles

for dinner, but tonight with two packs of tuna as a special treat. I was in my tent by 6:20. By now, my feet were no longer hurting from the 20-plus miles of daily abuse but rather numb and tingling when I finally peeled off my Merrells. I also noticed my socks were wet even though my shoes stayed dry, which was something that was becoming commonplace every day. I really liked the Gore-Tex boots for keeping water out, but after a grueling 12-hour day, my sweaty old feet stayed sweaty. And old. Tomorrow I would have just 11.9 miles to go to arrive at Trout Run parking lot at Swatara Gap by noon. My friend Bob had texted me from Delaware Water Gap to say he had just finished hiking all of Pennsylvania. He would be heading home to Pittsburgh tomorrow, so on the way back he planned to join David and Mom at the finish line.

Day 149
21.7 miles
Last night on the trail!

Day 150 Tuesday, November 17, 2020
Pilger Run Spring Trail (Appleby Campsite), PA to Swatara Gap, PA 72, PA

A 3:30 AM wake up, my last early-morning wake up on the AT! It was a nice clear, cool morning and I was on the trail by 5:10 with my trusty headlamp shining the way until 6:45. As I moved along the ridgeline, I passed by many amazing overlooks where I could peer down at the twinkling lights in the Lehigh Valley below me. It was easy travel this morning, as the path followed many long forest roads with flat cleared surfaces. It was almost as if I was being given a reprieve because it was my last day on the trail. I stopped for a few moments to gander at Round Top view in late morning; it was, of course, spectacular. I managed my time and mileage so as to arrive at noon at the designated meeting spot. As I followed the final ridgeline toward the gap, it seemed to be much longer than I expected from looking at the map. I kept plodding along, and finally there was a break in slope, and the trail began a steeper descent.

It was looking like I might arrive about an hour early, so I slowed my pace and began taking a few more short breaks. I arrived at the bottom of Blue Mountain and crossed under I-81. There, I saw a state-park tractor operator mowing the high, brown grass along Swatara Creek, which seemed odd to me at this time of the year. I followed blazes to an old bridge across Swatara River. It looked to me like an old railroad bridge, but it appeared to be positioned perpendicular to the old tracks, which had now been converted to trails. Later research revealed the bridge, built in 1890, was moved here in 1985 from its original location in Lycoming County. After crossing, I somehow expected a nice flat grade to the finish line. I was surprised when the trail suddenly began an abrupt uphill grade right after the river; this turned out to be where the trail rose up to cross the remnant of a nearly eroded ridge held up in part by Oriskany Sandstone. What could be a more appropriate way for Old Oriskany to end his 5-month adventure?

After another mile I noticed someone approaching in a blaze-yellow vest. As I got closer, I could see it was Bob, my old friend and fellow geologist. He made it as planned! We both hiked back to the parking lot together where I arrived right on the nose at noon. David and Mom arrived just 10 minutes earlier and were both elated to see me. It was a great day! They brought two giant #4 Jersey Mike's subs (provolone, prosciuttini and cappacuolo), apples, iced tea, and an assortment of snacks. For the next hour, we all exchanged questions and stories while feasting on this manna from heaven. It was so great to finally be finished!

Since starting back in early June, it's been a wonderful five-and-a-half-month journey through the diverse mountainous provinces of the Appalachian wilderness. This trail led me through places I always heard about and through places I never knew existed. I spent every hour of every day—from sunup to sundown—taking in the essence of this storied walk. My trek was far more than staring down at the trail each moment of every day, on the lookout for rocks, roots, and rattlesnakes. It opened my eyes to the grandeur and geologic past of this continent on which we all so briefly exist. This was indeed an experience of a lifetime that I will always cherish and remember.

Day 150
11.8 miles

The start of the 100-Mile Wilderness, Maine

View of Mt. Katahdin from Rainbow Ledges, five miles into the 100-Mile Wilderness

Trail Magic at Jo-May Road. Trail angel Pineapple prepared sausage and French toast with maple syrup and blueberry preserves for my unexpected second breakfast!

Overlooking Lake Onowa from Barren Mountain, 100-Mile Wilderness, Maine

Carnivorous pitcher plants in bog just south of Fourth Mountain in the Chairback Range

Blueberry-pancake breakfast at Shaws with other hikers, including Crazy Quilt and barefoot 4,000-foot summit hiker from Connecticut

Pre-dawn at lakeshore of Bald Mountain Pond, Maine

Early morning ascent to one of the several false summits on way to Bald Mountain, Maine

Eastern view into rising sun over Bald Mountain Pond from the summit of Bald Mountain, Maine

Looking east to sunrise shrouded in clouds over Avery Peak, while standing on summit of West Bigelow Peak, Maine

Morning Trail along ridgeline on Saddleback Mountain, Maine

Magnificent sunrise view from Bemis Mountain, Maine

Coming down Old Blue Mtn. into Black Brook Notch, looking southeast toward Partridge Peak and Ellis Pond, Maine

Navigating the boulder fields within Mahoosuc Notch, Maine

Another milestone!

Atop Wildcat "D" Peak, looking down into Pinkham Notch, NH

Hiking along the Carter Range with Fenway toward Mt. Moriah, Maine

View from Carter Notch Hut, looking south to Wildcat Mountain and its glacially scoured north wall of Carter Notch

Sunrise hike 3,000 vertical feet from Osgood Camp to summit of Mt. Madison, NH

Hiking along the Gulfside Trail in the Presidential Range nearing Mt. Washington, NH

Cog railway up to Mt. Washington, NH

A view from Webster Cliffs into Crawford Notch, NH

Trail leading up to Mt. Garfield Shelter, NH
Photo credit: Lionheart

Finally coming off of a three hour scramble up and over the ice-glazed Mt. Lafayette, NH

Looking back toward snow-and-ice-covered Mt. Lafayette on the following day from Lincoln, NH

Beaver Brook cascading over metamorphic ledges next to trail leading up to Mt. Moosilauke, NH

Fenway and me at Mt. Moosilauke summit
Photo credit: Emily Guay

Early morning sun reflecting off of beaver pond along Pressey Brook, NH

Scenic pasture along trail near Cloudland Road, VT

A rare evening campfire while tenting near Clarendon "Lodge" Shelter

Trail angel Ryan on top of White Rocks, VT

Looking like I lost too much weight! On trail in Green Mtns, VT

Great Falls, CT on the Housatonic River trail

In North Adams, MA with Dr. Dave, my son Josh's old Centre College roommate

In Berkshires, MA at Kay Shelter campfire with AT winter hiker Jason

Beaver pond next to trail south of Dalton, MA

Standing in a snowfield before heading into the Housatonic River Valley near Kent, CT

Nuclear Lake, NY

Hiking with Bob in NY on Bellvale Mountain

Dinner with Pittsburgh friends visiting for weekend: Bob, Robert, Steve, and me

Beautiful sunrise over fog-filled valley from Sunrise Mountain, NJ

Trail runner and FKT record holder for trail running NJ AT, "Quick with a K"

Crossing Lehigh River to my home county of Lehigh

Sunrise over Lehigh Valley from Bake Oven Knob, PA

With Mom at Rt 309 near Allentown, PA

With brother David at Rt 309, near Allentown, PA

Looking east toward the AT on Kittatinny Mountain from the Pinnacle, PA

Finished! After 150 days and just under 2,200 miles. At Swatara Gap, PA

Epilogue

So why did I decide to partake in this extreme form of "sufferfest," as a ski buddy recently described it, over a period of more than five months, feeling every day like I was either training for the Olympics or having enlisted in some sort of boot camp? There are a lot of reasons and stories to go with this explanation, and I'll do my best to make sense of it for the two or three of you who might still be reading. First, I have always loved the outdoors and that, in a big way, is why I became a geologist. I recall, on early family vacations, coming home with a rock-and-mineral set following a visit to a New England quarry town, or some other kind of geological attraction. I remember taking my first geology class as a freshman at Catawba College, after which I immediately changed my major from business (what was I thinking?) to geology, and never looked back! My fascination with the Earth and its natural processes over incomprehensible timespans has always been intoxicating to me.

Last year I spent a few months doing research on my family history and completed several lengthy and robust genealogy reports tracing at least four lineages back to Germany in the early 1700s. In doing so, I had the opportunity to learn a lot about past generations over the last three centuries. One of the things that I already knew, but certainly learned a lot more about, was my maternal grandfather and his siblings. I was intrigued to discover how so much of their lives were spent in the outdoors hunting, swimming, fishing or just rambling through the woods. Hundreds of old photos along with documented life events and old family stories really brought this to life for me.

Another thing I discovered in doing this research had to do with longevity. On my father's side, there has sadly been a history of relatively short lifespans for the men, ending at year 68, 68, 61, 87, 33, 64, and 79, in order, from my father to my 5x great-grandfather. Other than my 2x and 5x great-grandfathers (James Henry Eckert and Johann Justus Eckhard), most of these men died in their 60's or earlier. Already at 63, I'm sure with better nutrition, healthier living practices etc., I can beat this trend, but the stats are not in my favor. So, although it may seem irrational (and I hope it is), I wanted to spend time doing many of the things that I always had desired to do, like hiking the AT (now this is beginning to sound desperate) while there is still time. A simple response to the question of why I did this hike could be, "If not now, when?"

As a youth, my father who was an Eagle Scout, spent a great deal of his childhood, well into his teen years, hiking and camping all over Pennsylvania. He was even the bugler each year for Camp Trexler, our local Boy Scout camp in the Poconos. His passion for hiking and camping clearly rubbed off on my brother and me. My mother led a Girl Scout troop for many years and was equally committed to the scouting program. Growing up, I spent a lot of time with my Boy Scout troop either going on weekend backpacking trips, or short hikes frequently through the summer, many of which were along the Appalachian Trail through Pennsylvania. I probably hiked much of the trail through Pennsylvania more times than I can remember. Our family also spent time each summer camping, in one of those old canvas "heavy as a house," house tents. In addition, four friends from my church group who also had a passion for hiking actually did hike the entirety of the Appalachian Trail one summer in high school. I always regretted not having gone with Joey, Guy, Carol, and Debbie, and so I guess you could say that each of these factors have somehow combined to urge me on and give me a greater incentive to take on such a challenge. I also recently retired from a 38-year career as a geoscientist in the energy industry, a career that I pursued passionately. However, in that pursuit, I had spent more hours per week than I wished to continue. Having now more free time than ever before has finally allowed me to indulge in such extracurricular activities as this.

As revealed in the title of this book, the "Rocks" part was always in plain view and on my mind. It was fun and rewarding afterwards, combing through geologic maps and reports learning, confirming, and discovering lithological and structural aspects of the rocks about which I had spent so many months pondering. As a "soft rock" geologist (I mainly study sedimentary rocks), it was a relearning experience to get back to understanding the complexities of "hard rocks" (igneous and metamorphic rocks).

The path I followed on the AT was just one route of many through this storied chain of ancient, eastern mountains. I look forward to delving deeper into many more of the hidden gems lying out there to be explored. I recently spent four days in the Linville Canyon of western North Carolina exploring the rocky gorge and its many elusive trails. I was amazed at the geology and resulting physiography where the Linville Falls Thrust Fault, which formed the Grandfather Mountain Window, left behind an impressive escarpment of more resistant, cliff-forming Chilhowie Group quartzite units rimming the gorge.

Another trek I completed this summer with my old friend and fellow geologist Bob Heim, and his youngest son Ben, was the 70-mile Laurel Mountain Hiking Trail of western Pennsylvania. This trail essentially follows the ridgeline of Laurel Mountain, beginning at the Conemaugh River Gorge near Johnstown to the Youghiogheny River Gorge at Ohiopyle. Laurel Ridge is the structural expression of the Laurel Hill Anticline, one of the primary Salina cored anticlines within Pennsylvania's high plateau province. It was fun hiking over many of the geologic structures, which I have spent hundreds of past hours mapping from interpreted seismic and old gas-well data. Additionally, our hike led us through Seven Springs Ski Resort, which although I had been to a hundred times, I had never seen without snow!

In July I once again met up with Bob and his eldest son Nate, and we hiked together for the first 100 miles of the Long Trail (LT) in Vermont. At Willard Gap, Bob went east toward New Hampshire, and I continued north to finish the entirety of the LT (on July 28). The 273 miles took me deep into the wilderness of the northern sections of the nation's oldest continuous footpath, over such spectacular mountains as Camel's Hump, Mt. Mansfield, Belvidere Mountain, and Jay Peak. Of the 22 days I spent trekking over the rugged terrain of the Green Mountain Massif, only four of those days were without rain. On the days where visibility was afforded to me, the views were indescribably breathtaking.

Finally, in November I spent three days solo backpacking the North Fork Mountain Trail in eastern West Virginia. The entire trail follows the ridgeline of North Fork Mountain, which is the escarpment of the eroded southeast limb of Wills Mountain Anticline, held up by Silurian age Tuscarora Sandstone. This resulting escarpment produces a cliff-walk for much of the extent of this 25-mile-long trail. Along this trail are a myriad of vistas looking west into the Cambro-Ordovician carbonates of Germany Valley and beyond to the River Knobs, Fore Knobs and the Allegheny Front. The River Knobs are the result of the near vertical to vertical northwest eroded limb of Wills Mountain. Seneca Rocks is probably the best-known exposure of this vertical northwest limb and can be seen from the trail at many overlooks.

Long-distance hiking is a great way to explore and discover more about the country and world in which we live. There are countless other well-established trails to hike on foot, such as:

- The Pacific Crest Trail (PCT) and the Continental Divide Trail (CDT), the 2nd and 3rd parts of the hiker's North American Triple Crown.
- A northern, and trans-Atlantic extension of the AT, known as the International AT, is a geologist's true paradise of revisiting Pangaea on foot. This trail begins at Katahdin, and continues north into Maritimes Canada, eventually jumping across the ocean to the British Isles, Scandinavia, and other coastal parts of western Europe.
- The 1,400-mile-long Buckeye Trail in Ohio
- The 4,600-mile-long North Country Trail from New York to North Dakota
- And so many others in the Adirondacks of New York, or the Highlands of western North Carolina, just to name a few. Of course, each of these has a great geologic story to be revealed.

So many miles of trail, so little time...

Appendix

For most of my hike, my pack weighed between 35 and 45 lbs. depending on how much food and water I was carrying. Before changing out my pack, I was well over 50 lbs., when I started. A typical resupply added 10 to 12 lbs. to my pack weight. Each liter of water weighs 2.2 lbs., so carrying two full liters was an extra 4.4 lbs. I tried to use as much ultra light gear as possible, such as my tent (14 oz.), pack (2 lbs. 10 oz.), sleeping bag (1 lb. 3 oz. / 2 lbs. 4 oz.*), pad (1 lb.), and cook system (15.7 oz., gross wt.). A minimum of clothes was also important to reduce weight. I started out with a 6+ lb. L.L.Bean pack, and after 470 miles, changed to my 2 lbs. 10 oz. REI bag, losing over three lbs. At that point I also changed out my 20° down L.L.Bean bag for a lighter weight Big Agnes liner bag, losing about one lb.

My Gear

- REI Flash 55 Backpack (2 lbs. 10 oz.)
- Waterproof pack cover
- Pair of Trekking poles
- Sea to Summit waterproof bag for stowing sleeping bag
- Ditty bag for clothing
- Ditty bag for utility/first aid items
- Ditty Bag for cook pot, stove, fuel, etc.
- Rough Enough electronics bag
- Chum's surf shorts wallet
- First aid/toiletry items:
  - Toothbrush/toothpaste (keep in food bag)
  - Small comb
  - Several Dozen 1" x 3" waterproof Band-Aids
  - Adhesive tape
  - Ace bandage
  - Sunscreen
  - Lip balm
  - Leuko Tape
  - Small medical scissors
  - Tweezers
  - Fingernail clipper
  - Toenail clipper
  - Small tube antiseptic cream
  - Full Dose of antibiotics (e.g., Cipro)
  - Ibuprofen (Vitamin I)
  - Lightweight elastic knee brace

- - Ben's Bug spray, 100% DEET, 37 ml (optional, I rarely used it)
- Zpacks Altaplex 1-person tent (14 oz.)
- Tyvek Goundcloth
- 7 aluminum tent stakes
- L.L.Bean 20° goose down fill sleeping bag (2lbs. 4 oz. *early high alt. and fall only)
- Big Agnes Farrington sleeping bag liner (1 lb. 3 oz. *summer bag only)
- Outdoor Vitals Ultralight Stretch Pillow
- Sawyer Permethrin Pump Spray—24 oz. (treat clothing pre-trip for ticks)
- Therm-a-Rest NeoAir XLite Sleeping Pad (1 lb.)
- Therm-a-Rest Z-Seat Pad
- Zpacks Large Dyneema Food Bag (3.4 oz. empty, 10–12 lbs. full)
- MSR Pocket Rocket Stove and 8 oz. gas (2.6 oz. stove, 13.1 gross oz. fuel)
- Lixada Titanium 650ml Cook pot (2.8 oz.)
- Snow Peak HotLips
- TOAKS titanium Spork
- 2 Lighters
- Small pocketknife
- Small 4 oz. biodegradable liquid camp soap
- 4Monster Microfiber Towel (X-small size 12" x 20" hung on pack)
- Sawyer Squeeze Water filter system
- Water collection bags
- Potable Aqua Water Purification Tablets (iodine)
- 2 1-liter Smart Water bottles (4.4 lbs.)
- Toilet Paper
- Lightweight hand shovel
- Small 4 oz. hand sanitizer-hang from pack (refilled at each hostel)
- NatGeo Appalachian Trail Map Guides (1501–1513)
- ATC 2020 Appalachian Trail Thru-Hiker's Companion
- Small notebook for journaling and notes
- 2 Pens
- Cell phone (mainly for pictures, GAIA app for GPS guidance when needed)
- USB wall charger and cord
- Pocket Juice Battery charger (3.7V/8000mAh/29.6Wh, for iPhone and headlamp)
- Several dozen gallon zip-lock bags (for food, clothes, and any small items)

- Clothing:
  - Merrell Moab 2 Ventilator low cut hiking shoes (4 pairs over 150 days)
  - Altra Superiors as camp and water shoes
  - Dirty Girl Gaiters
  - Lightweight Gym shorts for night
  - Hiking Pants-Convertible Quick Dry Lightweight Zip Offs
  - Long sleeve silk long john top
  - Sleeveless polyester shirt
  - Puffy coat (polyfill insulated jacket)
  - Patagonia fleece pullover (lightweight)
  - Outdoor Research Rain Jacket
  - Marmot Gore-Tex waterproof rain jacket (purchased mid-Autumn)
  - REI Gore-Tex waterproof long pants (purchased mid-Autumn)
  - 3 pairs Smart Wool socks
  - 2 pair Injinji toe socks
  - Golf hat
  - Silk lightweight glove liners
  - Polyester gloves
  - REI Minimalist GTX waterproof mittens (shell) (purchased mid-Autumn)
  - Wool Cap
  - Buff or Neck gaiter
  - Sunglasses

Food

(5–6-day average re-supply, 10–12 lbs.)

- 12 packets instant Quaker oatmeal
- 2 boxes of 7 packets instant Folgers coffee
- 25 + assorted energy bars (Clif, Nature Valley, Larabar, etc.)
- 6 small boxes raisins
- 6 cheese or PB snack cracker packs
- 6 large beef sticks, Slim Jims or jerky
- 12 cheese sticks
- 6 Snickers or Almond Joy bars
- 6 assorted Ramen noodle packs
- 6 foil packs of tuna, chicken or spam
- 12 flavored energy mix packets for water

Trail Stats

Trail Time and distance Stats:

Measured quantity*	value	unit
Total logged miles:	2,157.2	miles
Average daily miles:	15.4	miles
Longest daily miles:	25.3	miles
Average steps per day	36,696	steps
Average steps per mile	2,576	steps
Average stride length	2.03	feet/step
Average speed	1.61	miles/hour
Average pace	37.3	minutes/mile
Average wake up time	5:08 AM	
Average trail start time	7:12 AM	
Average trail end time	4:37 PM	
Average hours per day on trail	9:40	

*Average values only include days where miles were logged on trail. ZERO days omitted. Steps per day from iPhone app. Only days on trail (when cell remained on all day) were used. Time averages only for those days I recorded times. Entire hours in a day used to calculate speed and pace (no break times were omitted, my actual average pace while hiking was about 30 minutes per mile). Many of the graphs below show these relationships over time.

Below is a brief summary of where I camped or stayed each night:

- 103 Tent nights
- 19 Shelter nights
- 27 Hostel or Motel stays for rest and resupply

Shelters:

June 14, 2020	Siler Bald Shelter
June 21, 2020	Silers Bald Shelter
July 7, 2020	Moreland Gap Shelter
August 31, 2020	James Fry (Tagg Run) Shelter
September 3, 2020	Clarks Ferry Shelter
September 29, 2020	Full Goose Shelter

September 30, 2020	Gentian Pond Shelter
October 6, 2020	Ethan Pond Shelter
October 7, 2020	Garfield Ridge Shelter
October 12, 2020	Firewarden's Cabin
October 13, 2020	Trapper John Shelter
October 16, 2020	Stony Brook Shelter
October 19, 2020	Little Rock Pond Shelter
October 20, 2020	Bromley Shelter
October 24, 2020	Mark Noepel Lean-to
October 28, 2020	The Hemlocks Lean-to
November 1, 2020	Mt. Algo Lean-to
November 2, 2020	Wiley Shelter
November 12, 2020	Leroy A. Smith Shelter

Hostel or Motel Stays:

June 5, 2020	Len Foote Hike Inn, Dawsonville, GA
June 16–17, 2020	Nantahala Outdoor Ctr., Wesser, NC
June 28, 2020	Laughing Heart Hostel, Hot Springs, NC
July 12–13, 2020	Broken Fiddle Hostel, Damascus, VA
July 29, 2020	Bear Garden Hostel, Ceres, VA
August 3, 2020	Angels Rest Hostel, Pearisburg, VA
August 15, 2020	Stanimal's Hostel, Waynesboro, VA
August 21–22, 2020	Motel, Luray, VA
August 25, 2020	Scott's Cabin, Snickers Gap, VA
September 11, 2020	AT Lodge, Millinocket, ME
September 15, 2020	Shaw's Hostel, Monson, ME
September 25, 2020	Saddleback Inn, Rangeley, ME
October 1–3, 2020	Rattle River Hostel, Gorham, ME
October 8, 2020	The Notch Hostel, N. Woodstock, NH
October 9, 2020	Liberty Lodging Motel, Lincoln, NH
October 14, 2020	Hanover AT Hostel, Norwich, VT
October 24, 2020	Willows Motel, South Adams, MA
October 29–30, 2020	Vanessa Breton House, Salisbury, CT
November 6–7, 2020	Motel, Sussex, NJ
November 11, 2020	Motel, Delaware Water Gap, PA

Expense Summary

I spent about $3,000 on all of my gear. Every five days (on the average) I spent about $65 on my food resupply. Over the course of 150 days I estimated that I spent close to $2,000 for food. Transportation by air to Atlanta plus round trip from PIT to Tri-Cities cost about $400. For numerous shuttles to and from hostels, into towns for resupplies and general transport, I likely spent over $500. I stayed 27 nights in hostels or hotels costing me over $1,000. Restaurant meals while in towns likely totaled well over $1,000. Altogether the trip cost me in excess of $7,900 over a period of five net months, or 150 days. This figure represents the total of all of my gear, all of my meals, hostel and hotel stays, and anything else in between. The average daily cost for this period comes to about $53 per day. Money well spent, in my opinion.

Graphs:

1. Type of night stay vs. Elevation
   This graph shows a simple elevation profile from Springer Mountain, Georgia to Mt. Katahdin, Maine. I also added the location and elevation of each tent camp, shelter or hostel location. It was so nice camping in my tent during the summer months, but the graphic reveals more shelter nights as the fall rains and temps drove me inside.

2. Miles per day hiked vs. Elevation
   This figure is a simple way to compare the effect, if any, of elevation on my daily miles. Once I left North Carolina, my mileage really picked up. It is also clear the toll the Whites and Mahoosucs had on daily miles logged. Curves show both individual miles per day, and a 10-day running mean average.

3. Miles per day hiked vs. Trail Hours per Day
   Here is a comparison of miles per day against hours spent on trail each day. Also shown is the portion of remaining daylight hours (morning and afternoon/evening). The effect of the shortening of days can be seen, especially in the latter half of my thru-hike as a SOBO. As fall progressed, not only were the days more quickly losing daylight, but in the last several weeks I put often in more trail hours than there were daylight hours. Curves show both individual miles per day, and a 10-day running mean average. Trail hours per day are gross time periods, i.e., no breaks removed.

4. Miles per day hiked vs. iPhone Steps per Day
As expected, a good correlation exists between miles per day and steps per day, as per my iPhone health app. These steps were next used to calculate my average stride length, in feet per step. Steps taken at camp in AM or PM are included in daily totals and considered negligible.

5. Stride length: Feet per step (miles x 5,280)/iPhone steps vs. MPH
Using my iPhone health app, I was able to calculate my average daily stride length, in feet per step. This was done by simple converting daily miles to feet and dividing by daily steps. The three and seven day running mean average curves are very smooth and produce a nice low amplitude sinusoid. Plotting MPH with both three and seven day running mean average curves shows a similar trend. The difficulty in terrain is revealed by these two parameters, more so than by looking simply as miles per day. The Smokies kept me at a shorter stride, and as I moved into Virginia, it is evident that my stride increased, and I was able to move faster, with easier hiking conditions. The Whites and Mahoosucs were perhaps the most difficult and slow-going terrain, and the curves clearly show this.

6. Age and Rock Type vs. Stride length
Throughout my hike, I continued to wonder whether the variability in rock type could have a noticeable effect on my performance. I used several sources for ages and rock types, most notably the exhaustive work compiled by V. Collins Chew and published in his 1988 geologic guide to the AT, *Underfoot*. I adjusted his geologic age dates a bit to comply with more recent work found, among several other place, on stratigraphy.org. I transcribed the rock type from Chew's appendix (Geologic Trail Log) onto a spreadsheet and then added the remaining details he provided by way of numerous V-lookup functions in Excel. My graph shows age range of rock units where I finished up each day. It is possible that there were cases where I had been travelling on one type of rocks for most of the day, and then coming into camp I finished on entirely different type of rocks. In cases where I camped near streams or in lowland shelters this is likely. I also show whether the rocks are sedimentary, igneous or metamorphic. With all these factors considered, there does not appear to be any compelling correlations between my stride length, speed or pace with rock type. Perhaps at a higher level of detail, a better correlation could be demonstrated. As with all graphs and models, they are a good starting point for a discussion.

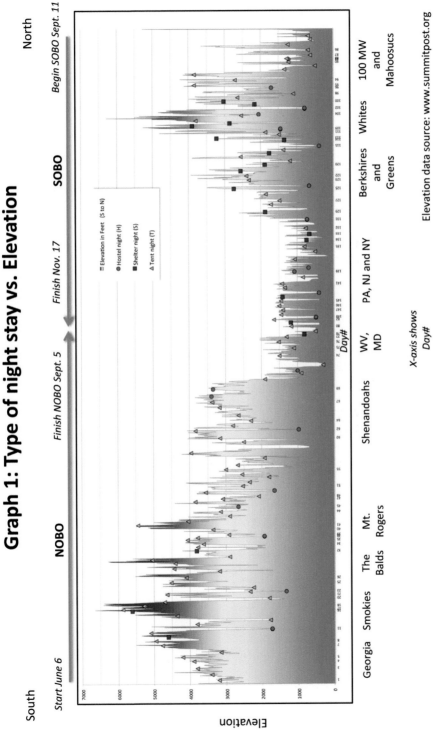

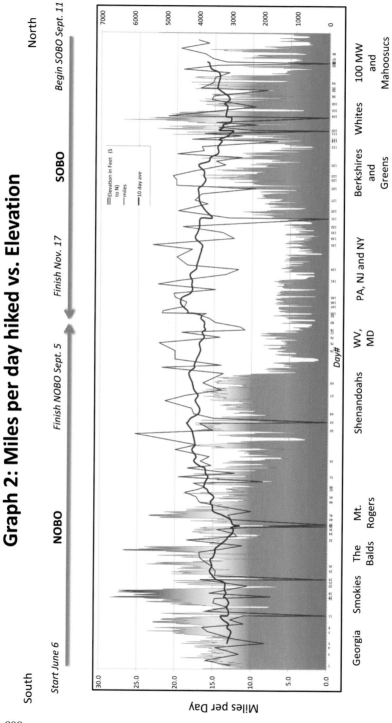

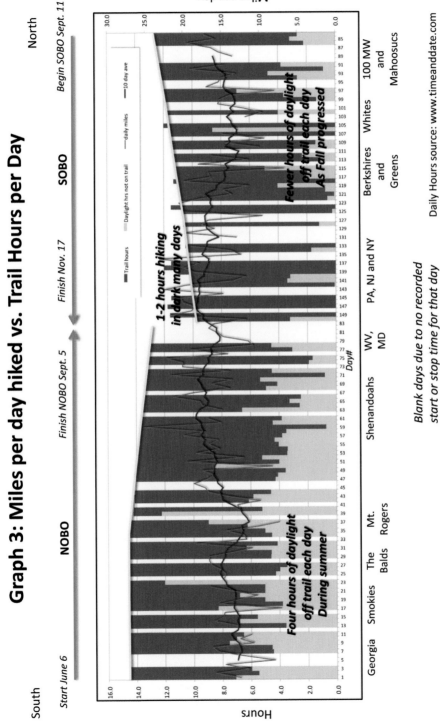

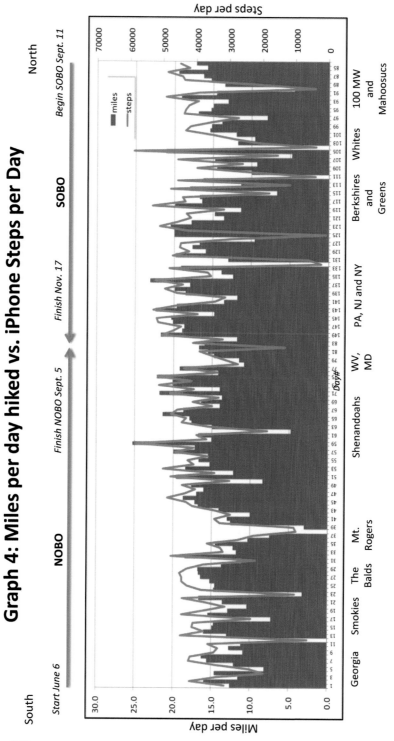

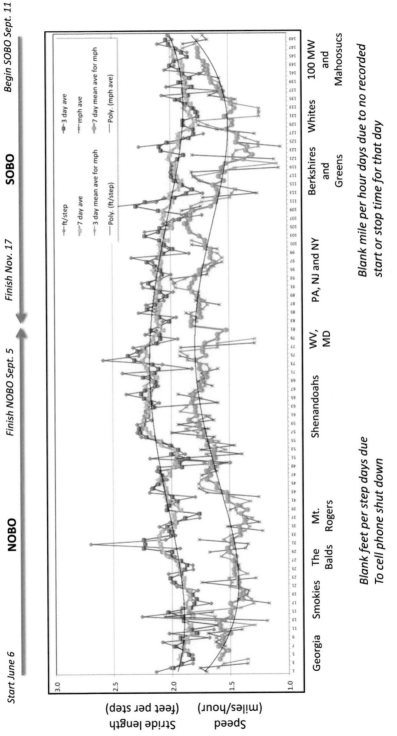

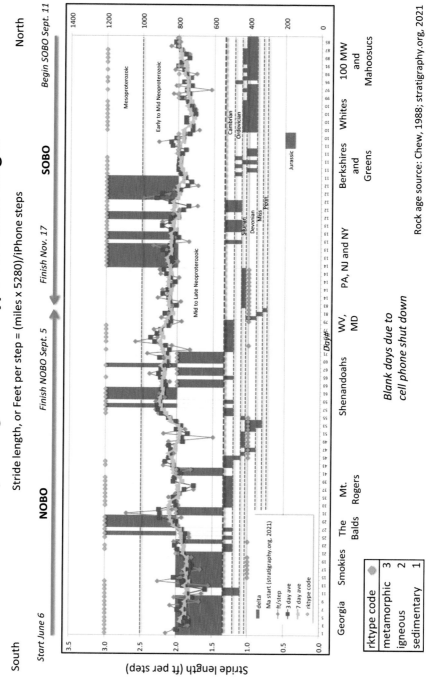

294

Selected References

Allison, I.S, Black, R.F., Dennison, J.M., Fahnestock, R.K., White, S.M., Geology: The Science of a Changing Earth, McGraw-Hill Book Company, New York, 1974

Berg, T.M., Edmunds, W.E., Geyer, A.R., Glover, A.D., Hoskins, D.M., MacLachlan, D.B., Root, S.I., Sevon, W.D., and Socolow, A.A., 1980, Geologic map of Pennsylvania (2nd ed.), Pennsylvania Geological Survey, Map 1, 1:250,000

Borns, Harold W. Jr.; Doner, Lisa A.; Dorion, Christopher C.; Jacobson, George L. Jr.; Kaplan, Michael R.; Kreutz, Karl J.; Lowell, Thomas V.; Thompson, Woodrow B.; and Weddle, Thomas K., "The Deglaciation of Maine, USA" (2004). Earth Science Faculty Scholarship. 276.

Cardwell, D.H., Erwin, R.B., and Woodward, H.P., 1968, Geologic map of West Virginia, West Virginia Geological and Economic Survey, Map 1, 1:250,000.

Chew, V. Collins, Underfoot: A Geologic Guide to the Appalachian Trail, (2nd ed.), Appalachian Trail Conference, Harpers Ferry, WV, 1988

Clark, S.H.B., 2008, Geology of the Southern Appalachian Mountains: U.S. Geological Survey Scientific Investigations Map 2830

Clark, S. H. B., Birth of the Mountains: The Geologic Story of the Southern Appalachian Mountains, US Department of Interior and USGS

Cleaves, E.T., Edwards, Jonathan, Jr., and Glaser, J.D., 1968, Geologic map of Maryland, Maryland Geological Survey, series unknown, 1:250,000.

Cook, Brian Stephen, "PALINSPASTIC RECONSTRUCTION AROUND A THRUST BELT RECESS: AN EXAMPLE FROM THE APPALACHIAN THRUST BELT IN NORTHWESTERN GEORGIA" (2010). University of Kentucky Doctoral Dissertations

Cutko, A., 2010, Natural Resource Inventory of the Mahoosuc and Rangeley Lakes Region: Mahoosucs, Maine Department of Agriculture, Conservation and Forestry, Bureau of Parks and Lands

Dalton, R. F., Monteverde, D. H., Sugarman, P. J., and Volkert, R. A., 2014, Bedrock geologic map of New Jersey, New Jersey Geological Survey, series unknown, 1:250,000.

Doll, C.G., Stewart, D.P., and MacClintock, Paul, 1970, Surficial Geologic Map of Vermont, Vermont Geological Survey, Miscellaneous Map MISCMAP-02, 1:250,000.

Doll, C.G., Cady, W.M., Thompson, J.B., and Billings, M.P., 1961, Centennial geologic map of Vermont, Vermont Geological Survey, Miscellaneous Map MISCMAP-01, 1:250,000.

Gallen, S.F., Wegmann, K.W., and Bohnenstiehl, D.R., 2013 Miocene rejuvenation of topographic relief in the southern Appalachians, GSA Today, v. 23, no. 2

Goldfarb, Ben, Eager: The Surprising Secret Life of Beavers and Why They Matter, Chelsea Green Publishing, White River Junction, VT., 2018

Hatcher, R. D., 2012, Blue Ridge Foothills Fieldtrip Guidebook, Teaching Structural Geology, Geophysics and Tectonics in the 21st Century Workshop, Tectonics and Structural Geology Research, Univ. of Tennessee-Knoxville

Hatcher, R. D., and Bream, B. R., 2012, New Tectonic Map of the Southern and Central Appalachians, scale 1:1,000,000.

Hatcher, R.D., Osberg, P.H., Drake, A.A., Robinson, Peter, and Thomas, W.A., 1990, Tectonic map of the U.S. Appalachians: Geological Society of America, Geology of North America v.F-2, scale 1:2,113,000.

Hardeman, W.D., Miller, R.A., and Swingle, G.D., 1966, Geologic map of Tennessee, Tennessee Division of Geology, series unknown, 1:250,000.

Hanson L. S. and Bradley, D. C., 1989, Sedimentary Facies and Tectonic Interpretation of the Lower Devonian Carrabassett Formation, North-Central Maine, in Studies in Maine Geology, Vol. 2, Maine Geological Survey

Horton, J.W. and Dicken, C.L., 2001, Preliminary digital geologic map of the Appalachian Piedmont and Blue Ridge, South Carolina segment, U.S. Geological Survey, Open-File Report OF-2001-298, 1:500,000.

Hibbard, J.P., van Stall, C.R., Rankin, D.W., and Williams, H., 2006: Lithotectonic Map of the Appalachian Orogen, Canada-United States of America: Geological Survey of Canada, Map 2096A, scale 1:1,500,000

Iwamori H., 1997, Heat sources and melting in subduction zones, JOURNAL OF GEOPHYSICAL RESEARCH, VOL. 102, NO. B7, PAGES 14,803-14,820

Lawton, D.E., Moye, F.J., Murray, J.B., O'Connor, B.J., Penley, H.M., Sandrock, G.S., Marsalis, W.E., Friddell, M.S., Hetrick, J.H.,

Huddlestun, P.F., Hunter, R.E., Mann, W.R., Martin, B.F., Pickering, S.M., Schneeberger, F.J., and Wilson, J.D., 1976, Geologic map of Georgia, Environmental Protection Division, Georgia Department of Natural Resources, series unknown, 1:500,000.

Lyons, J.B., Bothner, W.A., Moench, R.H., and Thompson, J.B., Jr., 1997, Bedrock geologic map of New Hampshire, U.S. Geological Survey, series unknown, 1:250,000.

North Carolina Geological Survey, 1985, Geologic map of North Carolina, North Carolina Geological Survey, series unknown, 1:500,000.

Osberg, P.H., Hussey, A.M., and Boone, G.M., 1985, Bedrock geologic map of Maine, Maine Geological Survey, Geologic Map Series BGMM, 1:500,000.

Overstreet, W.C. and Bell, Henry, 1965, The crystalline rocks of South Carolina, U.S. Geological Survey, Bulletin 1183, 1:500,000.

Prince, Philip S., and Henika, William S., 2018, Geology of the Catawba quadrangle, Virginia: Virginia Division of Geology and Mineral Resources Publication 187, 19 p.

Rickard, L.V., Isachsen, Y.W., and Fisher, D.W., 1970, Geologic map of New York, New York State Museum and Science Service, Map and Chart Series 15, 1:250,000.

Scotese, C.R., 2016, PALEOMAP PaleoAtlas for GPlates and the Paleo Data Plotter Program, PALEOMAP Project

Scotese, C.R., 2021. An Atlas of Paleogeographic Maps: The Seas Come In and the Seas Go Out, Annual Reviews of Earth and Planetary Sciences, v. 49, p. 669-718.

Sevon, W. D., Fleeger, G. M., and Shepps, V. C., 1999, Pennsylvania and the Ice Age (2nd ed.): Pennsylvania Geological Survey, 4th ser., Educational Series 6, 30 p.

Socolow, A.A., Geologic Map of the Appalachian Trail in Pennsylvania and Vicinity, 1983, Pennsylvania Topographic and Geologic Survey, Map 1, 1:250,000.

Sylvester, R., ed., Appalachian Trail Thru-Hiker's Companion 2020, Appalachian Trail Conservancy, Harpers Ferry, WV, 2020

Thigpen, J.R., and Hatcher, R.D., 2009, Geologic map of the Western Blue Ridge and portions of the Eastern Blue Ridge and Valley and Ridge Provinces in southeast Tennessee, southwest North Carolina, and northern Georgia: Geological Society of America, Map and Chart Series MCH097, scale 1:200,000.

Thompson, W.B. and Borns, H.W., Jr., 1985, Surficial geologic map of Maine, Maine Geological Survey, Geologic Map Series SGMM, 1:500,000.

Thornberry-Ehrlich, T. 2008. Great Smoky Mountains National Park Geologic Resource Evaluation Report. Natural Resource Report NPS/NRPC/GRD/NRR—2008/048. National Park Service, Denver, Colorado

Virginia Division of Mineral Resources, 1993, Geologic map of Virginia, Virginia Division of Mineral Resources, series unknown, 1:500,000.

Virginia Division of Mineral Resources, 2003, Digital representation of the 1993 geologic map of Virginia [The original 1993 map is available], Virginia Division of Mineral Resources, Publication 174, 1:500,000.

Woodward, H. P., 1932, Bull. 34, Geology and Mineral Resources of the Roanoke Area, Virginia, Virginia Geological Survey

Selected Websites:

https://ngmdb.usgs.gov/mapview/

https://appalachiantrail.org/

https://appalachiantrail.org/explore/hike-the-a-t/interactive-map/

https://blogs.agu.org/thefield/2019/04/29/the-landslide-that-is-too-big-to-notice/

http://hudsonvalleygeologist.blogspot.com/2014/12/nuclear-lake.html

https://www.mininghistoryassociation.org/Journal/MHJ-v23-2016-Kaas.pdf

https://www.summitpost.org/appalachian-trail-mileage-chart/593282

https://pubs.geoscienceworld.org/gsa/geosphere/article/3/4/220/31158/Tectonic-model-for-the-Proterozoic-growth-of-North

https://www.nj.gov/dep/njgs/enviroed/freedwn/WawayandaSP.pdf

https://www.nps.gov/grsm/learn/nature/geology.htm#10/35.6238/-83.5757

https://ngmdb.usgs.gov/Geolex/stratres/charts

https://www.usgs.gov/

https://earth.appstate.edu/facilities/southeastern-geology-digital-archive

http://www.nickzentner.com/

https://blogs.agu.org/mountainbeltway/

https://opengeology.org/historicalgeology/author/cbentley/

Endorsements

"This book is a great read. If you want to experience traveling the AT from your home, get this book. If you are the more adventurous, outdoors type, get this book and you will develop your own ideas for your next adventure. I did. Even if you are not a geologist, you will have some sense of the geology along the AT when you have finished the book. Being written in a journal format truly gives the reader the sense of being on the trail. After completing the book, I felt like I had been there with Craig. The pictures at the end of each of the three parts, helps the reader to visualize the terrain and personalities that Craig encountered on his journey. The analytics and gear list provide the basis for how to plan and execute your own trip. Overall, two thumbs up for a trip well planned, many lessons learned, and a book well written that should appeal to a wide range of readers."
Michael Canich, Golden Dog Geological Consultants

"Loved it! It feels like a conversation with you, which is a big compliment, your voice comes through! I was sort of sad to have it end, but as you noted, it was time to get off the trail."
Joan Crockett, Retired Petroleum Geologist, Illinois State Geological Survey

"If I had to describe it in just one word, I would say CAPTIVATING. I was hooked from the first page. I like your style—right to the point, sense of humor and education for those who are so inclined. If I were on vacation, I would not put this book down until finished."
Hana Kratochvilova, seasoned High Tatras hiker

"I thoroughly enjoyed following Craig Eckert's adventures and misadventures as he traversed the length of the Appalachian Trail. This book will appeal to all the Appalachian Trail thru-, day-, and armchair-hikers who have ever considered such an adventure. The journey is written as a diary with self-deprecating humor as he learns valuable lessons along the way for hiking and for life. Amongst his descriptions of the hikers and mountains he meets, his narrative intersperses comments about the geology that is underfoot—at times presenting him with difficulties, and at times with breath-taking views. His photos taken along the hike are spectacular and leave us wishing for more."
Dr. Robert Jacobi, Emeritus Professor of Geology, University at Buffalo

"Great read, great story, great adventure!"
Chuck Moyer, retired Geology Manager, Range Resources

"Many people dream and talk about it but very few find the resolve to do it. Hiking the Appalachian Trail is a hiker's bucket list item and Craig Eckert did it and tells an authentic tale of what it was like. The background on geology from an accomplished geologist adds to the tale. His informative story will inspire many to put on some boots and take the first steps on the Trail."
Neil Washburn, President, Material Advice, LLC